"The Object Lessons series achieves something very close to magic: the books take ordinary—even banal—objects and animate them with a rich history of invention, political struggle, science, and popular mythology. Filled with fascinating details and conveyed in sharp, accessible prose, the books make the everyday world come to life. Be warned: once you've read a few of these, you'll start walking around your house, picking up random objects, and musing aloud: 'I wonder what the story is behind this thing?'"

Steven Johnson, author of *Where Good Ideas Come From* and *How We Got to Now*

"Object Lessons describes themselves as 'short, beautiful books,' and to that, I'll say, amen. . . . If you read enough Object Lessons books, you'll fill your head with plenty of trivia to amaze and annoy your friends and loved ones—caution recommended on pontificating on the objects surrounding you. More importantly, though . . . they inspire us to take a second look at parts of the everyday that we've taken for granted. These are not so much lessons about the objects themselves, but opportunities for self-reflection and storytelling. They remind us that we are surrounded by a wondrous world, as long as we care to look."

John Warner, *The Chicago Tribune*

turns through which each of their authors has been put by his or her object. As for Benjamin, so for the authors of the series, the object predominates, sits squarely center stage, directs the action. The object decides the genre, the chronology, and the limits of the study. Accordingly, the author has to take her cue from the *thing* she chose or that chose her. The result is a wonderfully uneven series of books, each one a *thing* unto itself."

Julian Yates, *Los Angeles Review of Books*

The Object Lessons series has a beautifully simple premise. Each book or essay centers on a specific object. This can be mundane or unexpected, humorous or politically timely. Whatever the subject, these descriptions reveal the rich worlds hidden under the surface of things."

Christine Ro, *Book Riot*

. . . a sensibility somewhere between Roland Barthes and Wes Anderson."

Simon Reynolds, author of *Retromania: Pop Culture's Addiction to Its Own Past*

My favourite series of short pop culture books"

Zoomer magazine

Bloomsbury's Object Lessons series never misses"

The Millions

BOOKS IN THE SERIES

OBJECTLESSONS

A book series about the hidden lives of ordinary things.

Series Editors:

Ian Bogost and Christopher Schaberg

In association with

Lawn

GIOVANNI ALOI

B L O O M S B U R Y

NEW DELHI • LONDON • OXFORD • NEW YORK • SYDNEY

BLOOMSBURY INDIA
Bloomsbury Publishing India Pvt. Ltd
Second Floor, LSC Building No. 4, DDA Complex, Pocket C–6&7,
Vasant Kunj, New Delhi 110070

BLOOMSBURY, BLOOMSBURY ACADEMIC INDIA and the Diana logo are
trademarks of Bloomsbury Publishing Plc

First published in India 2025
First published in the United States of America 2025

Cover design by Alice Marwick
Cover illustration © Cristiano Siqueira / www.crisvector.com

This edition published with permission from Bloomsbury Publishing Plc,
50 Bedford Square, London, WC1B3DP, UK

British Library Cataloguing-in-Publication Data
A catalogue record for this book is available from the British Library.

ISBN: 979-87-65108-78-9

Series: Object Lessons

Typeset by Deanta Global Publishing Services, Chennai, India
Printed at Manipal Technologies Limited, Manipal

To find out more about our authors and books
visit www.bloomsbury.com and sign up for our newsletters

CONTENTS

PROLOGUE: TURF WARS

7:30 a.m. *What's with the noise outside?* I pull the curtain: two vans are parked in the neighbor's drive. Big boots and floppy hats—four workers offload heavy machinery.

Rototillers? Mowers?

Great! Finally, the Masons have caved in and called a landscaper to sort out that mess of a front yard. . . . I can't expect everyone to turn their yards into a pollinators' sanctuary, as I have done, but some basic TLC can go a long way. I gave them some tips last week: prune and water the boxwoods and the hardy hibiscuses that Lenny and Amy planted a few years ago—they look so tired. The lawn is a patchy mess. . . . *This might turn out to be good.*

I walk to the kitchen and get breakfast going.

Outside, things get loud. Not the way I like to start my day, but I remind myself that it is all for a good cause—a nicer neighbor's front yard!

Fig jam on toast is delightfully sweet. The smell of freshly brewed, dark roast coffee fills the kitchen. I focus on the bright side and get ready to work. *I have a busy day.*

9 a.m. First Zoom meeting with client.

11:30 a.m. Sexual harassment training. (*Is it the same cringy-clips every year or is it just me?*)

12:30 p.m. Staff meeting over lunch. *Ugh!*

2:00 p.m. Brainstorm/workshop session with the marketing team.

Will the landscapers' noise interfere with my meetings?

I pick up my laptop, ring-light, and charger. Mug in hand and folders underarm, I relocate to the bedroom on the opposite side of the house. I pull two chairs by the end of the bed— makeshift workstation: ready. I log into Zoom and push the dirty laundry out of the frame.

I ask my client if she can hear the rumble in the background. "No, not at all!" *Ah, technology—what a thing!* So, I carry on, one meeting after another. I only leave the bedroom briefly to grab a glass of water and pee.

By 11:30 a.m., the noise worsens.

What could possibly be happening out there?

Unable to contain my curiosity, I take a quick peek. A sod cutter sits in the driveway. The dead grass is completely gone.

This is radical!

The soil looks bare and desolate like a burnt old rug. *Great, I think—after years of neglect, the old grass was probably well beyond salvaging. A new beginning!* I raise my coffee cup to that and return to my bedroom-office. Things are quiet now . . .

. . . but not for long. The rhythmic thump-thump-thump of some kind of compacting machine kicks in. It gets louder and fainter.

Are they digging holes for new trees?

A chainsaw—that sound always gets my guts into a twist. *Some light pruning?* I'm glued to the screen—my presentation to the team is next in line. Not enough time to go back to the kitchen and take another quick peek to keep the growing anxiety at bay.

It's quiet again—the sound of hammering sporadically pierces the silence. Occasionally, a worker calls out to another.

I picture the neighbor's new front lawn. The new trees. A few new bushes? *Flowers, yes! Flowers.* I hope they kept some of the old shrubs too. A straggly juniper that, from a certain angle, looks a bit like an oversized bonsai looks cute. *Maybe they installed a water feature? A birdbath?*

First one, then the other: The vans rumble away.

The Zoom meeting drags on. Michael always has a last-minute question. . . . Another ten minutes . . .

Ah, it's over!

Urgently needed toilet stop.

Then the awaited moment of truth.

I pull the curtain

I squint my eyes.

My jaw drops.

Gone are the dead boxwoods and the spindly hibiscuses. The ground is green again, now covered by a perfectly even, uniformly emerald, and obnoxiously glistening AstroTurf carpet. Not only can I see it, but I can smell it too: a subtle but all-pervading stench of burnt tires fills my kitchen. Now everything is quiet. Very quiet. It couldn't be quieter.

A field of toxicity

Synthetic grass has been around since the 1960s. Invented by a team of researchers led by David Chaney, dean of the North Carolina State University College of Textiles, it was originally engineered to solve turf issues in sports fields. But more recently, due to increased and longer draughts, synthetic grass has become ubiquitous in residential areas too.[1] The artificial turf market is forecast to reach $7 billion by 2025.[2] Surprisingly, Europe, and not the United States, is leading the trend with landscaping businesses and the sports sector fueling a substantial increase in demand. There are many reasons behind the success of synthetic grass. Maintaining a lawn is labor-intensive, time-consuming, and expensive. Fertilizers, pesticides, and weed killers—lawns are endlessly thirsty. Mowers and leaf blowers are costly, noisy, and highly polluting.

Synthetic grass never yellows, and it is comparatively far less unpredictable. But despite the initial advantages, it might not be the long-term answer to the problems of patchy lawns it promises to be. A field of toxicity: the granular matter used in the infill of synthetic grass—the stuffing that makes the blades stand up—is made of shredded old tires: a deadly mix of heavy metals, cancer-causing compounds, and irritants linked to breathing ailments.[3] The artificial grass blades, often made from nylon, quickly heat up under the sun, making the lawn unwelcoming to humans as well as pets. Summer after summer, the shimmering green fades into a sickly bluish hue. And to make matters worse, synthetic lawns continually release invisible microplastics into the atmosphere and deep into the soil. Their impermeability destroys bacterial and fungal life underneath it.

While it might not require mowing, a synthetic lawn, like an indoor rug, needs hoovering and washing. It just can't adequately support biodegradation. Organic matter gets trapped between the fake blades and slowly rots—it stinks. While manufacturers claim that an artificial lawn will last twenty years, average synthetic grass greatly deteriorates after only seven.[4]

The popularity of synthetic lawns is only the tip of a massive, green iceberg. Laying green plastic carpets made of recycled car tires over already seriously compromised ecosystems is hardly what we need to remedy climate change and support local and relentlessly threatened wildlife. The synthetic lawn is a quintessential incarnation of our deep disconnect with nature—a monument to our alienation from other species and ecosystems with which we are carpeting the world. Ultimately, covering the land with artificially mass-produced reproductions of the traditional grass lawn is the ultimate commodified incarnation of the capitalist logic that has led us, via colonialism, to climate change in the first place.

To the root of the matter

Underneath the green blades of grass of every lawn, synthetic or not, lies a deep stratification of intricate ideological and ecological issues that, over time, have become naturalized. Where and when did the lawn originate? Why is its aesthetic allure so enduring? What does the lawn represent? How can we overcome our lawn addiction? What sustainable alternatives are at our disposal?

Over the last century, the lawn has taken over the world. NASA satellite imaging reveals that turf grasses cover 40,000 million acres of land in the United States alone—nearly the size of Washington State. This by far exceeds the *combined* acreage allocated to growing corn, wheat, and fruit orchards.[5] There was certainly a time, not so long ago when the lawn was almost universally admired as a status symbol. After the Second World War, lawns were fetishized in the United States as the quintessential incarnation of the American Dream. Neat, clean, manicured, perfect: the lawn embodied the very essence of modern, suburban masculine pride.[6] But by the beginning of the new millennium, the allure started to fade. Today the grass is no longer greener on either side of the fence. In fact, there's no green grass so far as the eye can see. This has now become the norm through the increasingly hot summers in both hemispheres' temperate zones. Consecutive heatwaves and dwindling rainfall have turned the once beloved, lush-green lawns into scorched and barren wastelands.[7]

It is, therefore, no surprise that the lawn is now at the center of a climate change controversy—the emblem of privilege, the enemy of biodiversity, and the death of ecology. The large carbon footprint maintenance, its unquenchable thirst for fertilizers, weedkillers, and water, and the notorious unfriendliness toward all forms of wildlife have attracted mounting criticism and even spurred an anti-lawn movement in the United States.[8]

But why did it become so popular in the first place? The lawn is just as aesthetically versatile as it is ideologically loaded. From museums and hospitals to corporate headquarters, university campuses, and cemeteries, it has become the *verdant lingua franca* of institutions of all kinds. Its formal homogeneity and neatness speak a universal vernacular of power: reliability, constancy, efficiency, confidence, and trust. The lawn is a complex code we decipher at a glance. It says so many things at once. In truth, most of what it conveys has nothing to do with nature and almost everything to do with us, our perceptions, ambitions, social anxieties, prejudices, illusions, hubris, and fears. Never just a soft carpet of grass, the lawn is a green mirror in which all we see is our reflection.

The lawn always is, first and foremost, a cultural field of intensities, discourses, power relations, genealogies, rhetoric, and ideologies. This explains why over the past fifty years, it has become the subject of many books—some in praise of man-made grassy fields and some thoroughly critical of them. Virginia Scott Jenkins's *The Lawn: A History of an American*

Obsession was published in 1994, and it is among the very first, modern critical books to explore the American lawn as a cultural icon, questioning its environmental sustainability and its role in promoting conformity and social inequality.[9] Jenkins's lucid analysis set the blueprint for the authors who followed in her footsteps. Among others, Tom Fort's *The Grass Is Greener* delves into broader questions about British identity, class, and history. While Georges Teyssot's *The American Lawn* is a valuable edited collection featuring contributions by Jenkins as well as many other women authors.[10] Still largely concerned with the historiographical/sociological path, the more recent *American Green: The Obsessive Quest for the Perfect Lawn* by Ted Steinberg and *Lawn People: How Grasses, Weeds, and Chemicals Make Us Who We Are* by Paul Robbins address environmental issues in more resolute ways and provide in-depth data analysis.[11] By far the most philosophically and academically grounded title published on this subject is Jonathan Cane's *Civilizing Grass* (2019). The book focuses on South Africa, instead of the recurring European, British, and US landscape, and it mobilizes a wide range of philosophical lenses ranging from LGBTQIA+ performativities and Marxism to Visual Culture.[12] Certainly worthy of note is the pioneering *Redesigning the American Lawn* by Herbert Boremann, Diana Balmori, and Gordon T. Geballe. Published in 1993, the book was the very first to thoroughly focus on environmentalism, proposing actual alternatives like the "freedom lawns": low-maintenance biodiverse meadows that reduce the need for chemical treatments while promoting resilient ecosystems that can better withstand pests and diseases.[13]

While all these books, in one way or another, inform this one, *Lawn* is radically different from them in structure and approach. This book is more philosophical than it is historically grounded.[14] It is deeply informed by key concepts derived from the fields of critical plant studies,[15] posthumanism,[16] speculative realism,[17] object-oriented ontology,[18] indigenous knowledge,[19] and queer ecologies,[20] among others.

While never losing sight of the vegetal essence of the lawn, this book constantly moves between the immanence of mowers, blowers, gasoline fumes and weedkillers, and the transcendence of symbolic registers that have turned the lawn into an essential cultural icon as well as an ecological catastrophe of our time. Ultimately, the lawn is a spatial entity across and through which power relations, labor, networks of signification, aesthetic expressions, and climatic manifestations can be visualized and deconstructed. Climate change has made it clear that the cultural forces that define our relationship with the lawn must be dissolved in the prioritization of interspecies gatherings, cominglings, coexisting, and coevolutions.

But first and foremost, I hope this book will instil a sense of responsibility. For those of us who own lawns, work for institutions that do, or can muster the courage to approach local governments about the need to reimagine public green spaces, this book proposes a range of experimental approaches to reimagine human-nature partnerships and lay stronger foundations to support more sustainable futures.

This book is structured around the quintessential spaces that have become central to the different iterations of the lawn in our time: grasslands, sports pitches, front lawns, backyards, parks, golf courses, buffer zones, and gardens. Regardless of whether these lawns are composed of the same kind of grass and might at first glance seem similar, their symbolic significance and the power dynamics that define them vary according to their function and the social roles they are expected to play. These grassy zones of interactions define our present and outline our future, locking us in or out, entrapping our thinking, and stunting or expanding our range of action. Mapping the power relations involved in our relationships with different kinds of lawns is the first step toward productive regeneration.

Chapter 1, "Grasslands," presents a precapitalist vision of grassy fields by enmeshing biological and ecological richness and complexity with historical memories, myths, and narratives. It invites us to reconsider the cultural emphasis we currently place on forests as quintessential models of ecological harmony in contemporary discourses to think *with* the decentered multiplicities of grasslands instead. Rather than emphasizing the anthropocentric nature of trees, this chapter explores how to negotiate diversity and unpredictability, mindfulness and patience, struggle and reinvention in new and challenging ways.

From Dante Alighieri's interpretation of Elysium to the Gardens of Versailles, the second chapter, "Yards," traces a global history of the lawn as the indispensable virtue-signalling status symbol we have inherited from Old World aristocracy. In line with the moral standards of the Enlightenment, the sophistication of one's own education and manners should be reflected in the refinement of their material possessions. During the first half of the twentieth century, this long-held belief was revitalized as the lawn began to incarnate the ideal of a new modern man—the keeper of the "good and proper"—by fostering a performative reconnection with a tame nature that he could successfully maintain with relative ease. This reconnection bolstered threatened masculinity. What cultural forces shape our relationship with the front lawn today, and how are these related and yet different to the ones that configure our engagement with backyard turf?

During the 1920s and 1930s, the patriarchal ideals of health, strength, heterosexuality, and recreation—fundamental to team sports like football, rugby, and cricket—became central to "happy modern living."[21] Chapter 3, "Pitches," focuses on the genealogical analogies between field sports like baseball, rugby, football, and military training. From this vantage point, sports turf is revealed as a metaphor of the battlefield. Although groomed and tamed, the pitch is a material abstraction of the land upon which deadly confrontations unravelled centuries ago.

By the end of the eighteenth century, as the industrial revolution propelled the rise of a new mercantile class, closely shorn carpets of grass had become extremely fashionable across Europe. It was against the background of sprawling urbanization and an unprecedented alienation from nature that affordable push lawnmowers and garden hoses marked the beginning of a new chapter in the story of the lawn. Under the dichotomous structure of humanist philosophy, the unwieldy entanglements of weeds started to metaphorically stand for queerness, while the homogeneity and pragmatism of the grass embodied the ideals of modern masculinity.

Building upon the previous, Chapter 4, "Parks," contextualizes the lawn as an essential and normative panoptic scopic field. In the history of Western positivism, visibility is a powerful, normative force.[22] It holds in place; it prevents, it deters; it encodes; it enables identification and classification. It exposes to shame, but it also shelters from violence. The park lawn is uncompromising in its flickering aesthetic integrity—a quasi-fascist bundle of bourgeoise qualities rooted in patriarchal ideals. Green grass, with its implied promise of stable futurity and sustenance, holds in place the stereotyped and homogenized image of a repressed social order.

Chapter 5, "Golf Courses," maps the recent proliferation of one of the most dreadfully special kinds of turf in the world. A living emblem of supreme capitalist commodification, more than any other kind of lawn, golf turf is in every sense *hyper.* A thoroughly "commodified pastoral," its vast expanse induces a kind of aesthetic experience imbued with a deadly, imperturbable serenity. The greenest grass of them all: golf courses notoriously are where important business deals are made. Primarily a picture; golf lawn aesthetic values are grounded in the apparent naturalization of a capitalist social order that keeps corporations in power. Like a plush carpet concealing the basic struggle, pollution, exploitation of lower social groups, and the environment, with its promise of a vapid and eternally stable aesthetic serenity, the lawn hides corporate avarice. What genealogies and ideological analogies entangle the golf turf to the lawns of corporate parks and to those that surround university campuses?

Road verges, highway shoulders, roundabouts—12 million acres of roadsides and median strips across the United States are regularly mowed, unnecessarily polluting the planet and further reducing opportunities for wildlife to thrive. Perhaps more than any other application of the lawn in contemporary times, these short-shorn swaths of grass exemplify the cultural pervasiveness of the lawn as more than a pragmatic solution to green management. Chapter 6, "From the Buffers, Back to the Garden," focuses on often neglected, liminal spaces like highway medians as opportunities to rethink our relationship with lawns and the land. An "omnipresent planetary entity," as Reza Negarestani would call it, the lawn has become an active and powerful biological, as well as political, agent, its roots reaching deep into our psyche, into our fears, anxieties, and misplaced, utopian hopes.[23] Have we domesticated grass or has grass domesticated us?

Artists show us that urgent and effective actions are needed to mitigate the negative impacts of lawns across the globe as giving up the lawn for good is not as easy a task as initiatives like No-Mow May suggest. It is now evident that we must participate in a longer-term process of aesthetic and intellectual re-education in order to wean ourselves from our lawn addiction. Our responsibility to care for our yards now extends beyond the well-being of our family—the pollinators, the water, the soil, the air, and the invisible networks of fungi and bacteria that support all life on this planet. In fact, no yard is too small to matter. My hope is that this book will inspire readers to take action and change the world, starting from their own front or backyard, one lawn at a time.

1 GRASSLANDS

Covenants of reciprocity

Look up into the Eastern sky at dawn or search the Western horizon at sunset, just before the first stars appear, and you may see Venus, our solar system's brightest celestial body after the sun and moon. To the Yolngu tribe of Australia's north-eastern Arnhem Land region, Venus is a creator spirit: low in the sky, it whispers to the earth. To the peoples of the Hamilton River, a tributary of the Georgina River in Queensland, Venus is called *Mumungooma*, meaning "big eye."[1] It is imagined as a fertile grassland—a provider of nutritious seeds, leaves, and fruits.[2] The recurrence of grasslands as an otherworldly space of abundance and generosity in Aboriginal mythologies is a testament to their ecological and cultural importance. In opposition to common misconceptions, Indigenous peoples actively interacted and managed the land, mindfully using fire to prevent the expansion of forests as well as to regenerate the fertility of the soil. Incinerated tough and old shoots bestowed nourishment for new growth. For over 10,000 years, in the

Northern Great Plainsof America, throughout the seasons, bisons moved from the open spaces of prairies to the foothills along the Rocky Mountains in search of different grasses: *Hesperostipa comata* was a favorite in cold weather, whereas *Bouteloua gracilis* was a summer staple.[3] Apache, Arapaho, Blackfoot, Cheyenne, and other First Nations tribes relished an intimate understanding of the animals' connection to the land and how grasses guided their movements across it. Bisons tilled the soil so that seedlings could easily sprout. Knowledge of these complex relationships, which was frequently passed on by women who gathered and prepared medicines, was critical for survival. Indigenous tribes, animals, and lands thrived on this attuned interplay in which death never came at once but flickered as a regenerative force—never an end— among the fires, the floods, and the dry spells. These early land-stewardship practices might have been in use for over 10,000 years, thus deeply albeit slowly, carving the landscape through co-evolutional shifts.[4]

Across the globe, grasslands have always been synonymous with generosity. The biodiversity, the accessibility, the manageability, the sustenance. The resilience and longevity of grasses also stood out as exceptional qualities. In a myth crafted by the Mende of Sierra Leone and the Dagomba of Ghana, a goat interfered with the messenger's mission of a dog, stealing from it a pouch containing a sacred medicine concocted by the Krachi Creator Wulbari that would have granted immortality to humans. According to the tale, the goat scattered the pouch's medicine upon a grassland thus incidentally making it immortal.[5] Recurring in different versions of this tale as told by other local tribes, the connection between human mortality and grass immortality is not a little detail. Health and prosperity in many Indigenous communities could only be built on a humble sense of reciprocity and parsimony, not on reckless exploitation.[6]

Potawatomi scientist Robin Wall Kimmerer meaningfully illuminates these ever-evolving relations as a "covenant of reciprocity":

> plant breath for animal breath, winter and summer, predator and prey, grass and fire, night and day, living and dying. Water knows this, clouds know this. Soil and rocks know they are dancing in a continuous giveaway of making, unmaking, and making again the earth.[7]

But because of the soil richness and flatness, since the beginning of colonization grasslands have constantly been threatened by agricultural overtake. Illinois still calls itself the prairie state while less than 0.01 percent of its original 21 million acres of tallgrass prairie survive today.[8] With 80 percent of ancient grassland having been turned to mono-crop agriculture, the remainder of North America does not fare much better.[9] Over 90 percent of the grasslands in Europe have also been lost since the 1930s.[10] And more than 43 million hectares of the Eurasian steppe are also gone as is 60 to 80 percent of the grassland area in South America.[11]

In an effort to reverse the environmental deterioration that is hastening the sixth mass extinction and exacerbating climate change, scientists are more interested than ever in the idea of integrating Indigenous land stewardship methods with technological advancements. But at the core of this important and urgent plan lies an equally urgent issue that must be addressed. For any ecological remediation to truly work long term, we must first reeducate ourselves. The sense of interconnectedness nurtured by North American Indigenous cultures around the world was once also present in Europe and other parts of the world. Our alienation from nature is a relatively recent and very complex phenomenon caused by relentless ideological structures and sedimentations that can be dismantled and reconfigured.

The start of the Italian Renaissance in the second half of the fourteenth century marked a turning point in the evolution of humanity's alienation from the more-than-human world. In time, the Western idea of human universal centrality (anthropocentrism) was significantly bolstered by the Black Death, a worldwide pandemic that claimed an estimated 200 million lives in North Africa and Eurasia between 1347 and 1351. This catastrophic event forced mankind to grapple with a deep existential crisis of unparalleled magnitude.[12] Why would God inflict such a terrible punishment on his most valued creation? It is in the aftermath of this tragedy that a new philosophical current emerged: humanism—a formidable antidote. More than any other philosophy, humanism laid the foundations of today's Western world. For a while, it served the purpose: it gave humans self-confidence. But at what cost? The art of the Renaissance exudes positive grandeur, a sense of imperturbable stability and centrality that humanism so strongly promoted. Its inherent hubris served us well until the eighteenth century when it morphed into blatant arrogance. The most incredible architectural and artistic marvels; the flourishing of literature, music, and science; the transatlantic slave trade; the never-ending fighting between counties, cities, and kingdoms; the looting, burning, and raping; the awe-inspiring technological innovations—the sheer sense of protagonism that pervaded Western civilization: humanism, incessantly reaffirmed by Christianity, finally gave birth to a patriarchal world system that enabled imperialist crimes, genocide, relentless environmental devastation, racism, sexism, and everlasting conflict.[13]

Not only humanism forever changed the West, but over the 500 years that followed its emergence, it swallowed the rest of the world. Its ideological foundations quickly stratified and calcified in our minds, carving insurmountable canyons and giving rise to the peaks of neoliberal capitalism that are today responsible for the current ecological crisis. In *The Great Derangement*, Amitav Ghosh argues that the current capitalist system, with its reckless and inconsiderate pursuit of growth and profit, has created a culture of denial and indifference toward nature and climate change. He rightly believes that the capitalist system encourages short-term thinking and rampant individualism, which prevents us from recognizing the long-term consequences of our actions on the environment. This, he argues, is a form of "derangement" that makes us incapable of addressing the climate crisis with any true determination and efficiency.[14] Humanism is the root of this derangement.

Ecological activist Vandana Shiva has also been a vocal critic of the negative impacts of globalization and free-market capitalism on developing countries, particularly in relation to agriculture and food security. She argues that the current capitalist system is inherently unsustainable and exploitative, leading to environmental degradation, social inequality, and the erosion of Indigenous cultures and knowledge. Shiva contends that capitalist systems promote agricultural monocultures that reduce biodiversity, disrupt ecosystems, and increase vulnerability to pests and diseases. This, she believes, is harmful not only to the environment but also to the livelihoods of small farmers who are unable to compete with large corporations like Monsanto. Shiva's eco-capitalist critique thus focuses on the issue of intellectual property rights. She argues that capitalism's conceptual emphasis on private property extends to the realm of knowledge and biodiversity, leading to the patenting of life-forms and traditional knowledge. This, she asserts, is a form of *biopiracy* that dispossesses Indigenous communities of their resources and knowledge.[15] Intensive plant breeding, synthetic fertilization, and mechanization of processes have been a double-edged sword—on one hand, food is abundant and affordable for some but on the other, the planet, and us with it, is catastrophically impoverished.

The forms of cultural resistance enacted by authors like Ghosh and Shiva, among others, at times influence lawmakers and political leaders who can truly make a difference. But this is only the beginning of a long and truly indispensable process of reeducation about our relationship with plants, animals, and the land.

Beyond anthropomorphic roots

In 1999, botanists James H. Wandersee and Elisabeth E. Schussler theorized the concept of "plant blindness" to exemplify the condition that affects most of us who live in the West today. According to them, "plant blindness" is defined by

(a) the inability to see or notice the plants in one's environment; (b) the inability to recognize the importance of plants in the biosphere and in human affairs; (c) the inability to appreciate the aesthetic and unique biological features of the life forms that belong to the Plant Kingdom; and (d) the misguided anthropocentric ranking of plants as inferior to animals and thus, as unworthy of consideration.[16]

But what does it mean to be blind to plants since we can clearly see them everywhere around us? It means that cultural blinkers, like humanistic philosophies, religions, and capitalist ideologies, have instilled in us a narcissistic sense of superiority that makes us blind to the wonders of the natural world. We see a plant, but our gaze rarely gets past its surfaces, colors, and shapes. We can only perceive it as a beautiful object or a useful resource. We remain blind to its hidden sensorial complexity and behavioral sophistication. Hierarchies, notions of rarity, returns on investment: capitalism has nullified our ability to pay true attention to nature in deep and meaningful ways.

In this sense, grasslands are more vulnerable than other plant-based ecosystems. Part of the issue lies in a persistent form of aesthetic miseducation. Centuries of paintings, photographs, documentaries, and movies have trained our gaze to only relish the sublime elegance and grandness of charismatic megafauna and majestic trees. The exceptional, the regal, the monstrous—capitalism has distorted our perceptual bandwidth by warping our value systems and entrapping us in predictable economies of the spectacle. Lions, tigers, bears, elephants, sequoias, and baobabs—the infiniteness of nature's biodiversity is shrunk to a handful of commodifiable, charismatic species in which we only see reflections of ourselves.

It is no coincidence that today, contemporary cultural conversations on ecology should often promote a perilous fetishization of the forest as an implicit (and yet false) pre-agricultural and wholly benign ecosystemic model of pure nature.[17] A plethora of books and newspaper articles celebrate the complex ways trees communicate with each other through mycorrhizal networks of fungal symbiosis inviting us to consider them as viable models around which societal relations might be restructured. Bestsellers like Peter Wohlleben's *The Hidden Life of Trees* or Suzanne Simard's *The Mother Tree* intentionally mobilize highly anthropomorphic tropes to instil empathic responses from readers. These books might enthuse wide audiences, and they are informative, although they do not truly reeducate us deeply enough. Oftentimes, the authors deliberately turn to archetypal, normative, and conservative pre-packaged notions designed to reassure us that, if we look carefully, trees are like us. They don't dismantle the humanist structure that keeps us, humans, at the center of everything.

Anthropomorphism is a highly complex representational phenomenon that encompasses many mediatic forms of expression. It is an intrinsic part of our species' thinking structures and, in a sense, it is inescapable. Anthropomorphism is always deceptive for it imposes human form on other, radically different, earthlings. If escaping it completely might not be possible, responsibly finetuning and modulating the ways in which it produces knowledge is of paramount importance. For these reasons, in this book, I intentionally follow a more arduous path—one that constantly demands us to negotiate the challenges posed by diversity and unpredictability, mindfulness and patience, struggle and reinvention.

It is in this context that thinking *with* (and not simply about) grasslands presents a much more exciting and enriching challenge. Grasslands, in their precapitalistic form, are first and foremost all-encompassing, vibrant, constantly morphing, and highly diverse ecosystems of enormous importance to the life of our planet.[18] The 57 million acres of grasslands comprising the prairies of the Western Great Plains of North America support the lives of more than 600 species of birds like the lesser prairie-chicken, upland sandpiper, and thick-billed longspur as well as mammals like bison, black tail prairie dogs, black-footed ferrets, swift fox, burrowing owl, and mountain plover. At certain times of the year, grasslands are open-air granaries. And, of course, they are home to thousands of pollinator species from bees to wasps, butterflies, and moths. The much-celebrated forests of contemporary culture can't quite compete.

It is often recited that trees are the lungs of the earth, but grasses produce comparatively much more oxygen.[19] And since it mostly takes place below ground, much of their ecological contribution remains invisible to us. Grasslands are all about the intricacies of interconnectedness, from the entwining green blades that provide shelter and retain moisture to the extensive root system that generates an intricate web of interdependencies between the soil and its inhabitants. The carbon dioxide absorbed through the photosynthetic process is stored in the roots which are significantly longer than those of other plants, lawn grasses especially. The roots of the humble buffalo grass (*Buchloe dactyloides*), for instance, can reach as deep as nine feet.[20] While blades of grass protect the soil from wind and water erosion, the roots increase its fertility and permeability.[21] The soil greatly benefits from the *mycorrhizae*, a symbiotic relationship established between grasses and fungi. *Mycelia*, the fungus root network, gains access to carbohydrates directly from the roots of the plant and the plant increases its efficiency in the uptake of water and nutrients, thus boosting its metabolism and overall health and resilience to parasites and pathogens.[22]

However, while grasslands are such marvellously rich and vibrant ecosystemic realities, from an aesthetic standpoint, they pose many challenges to our modern disaffected gaze. Grasslands make anthropomorphism difficult. Paradoxically, despite their fullness, some types of grasslands are, from an aesthetic standpoint, perceptually close to deserts. Their flatness, openness, uniformity, and apparent visual homogeneity are instantly perceived by us as "vast, open spaces." Forests are not subjected to quite the same reductive perceptive tyranny. Their multiplicity is heterogeneous even when trees of the same species dominate; each has its distinctive form and bears its scars. Their verticality echoes that of our bodies. Their longevity frequently compels us to attribute a higher value to their lives. It's no coincidence that we plant a tree instead of a patch of grass to commemorate the birth or passing of a family member.

To appreciate the uniqueness of grasslands might seem akin to expressing sympathy for the individuality of desert sand grains: a pointless endeavor. Grasslands are always a plurality, an indistinguishable multitude, a mass that in our eyes far too swiftly dissolves into a color and a texture. This inescapable multiplicity can become a liability. But as artist Jenny Kendler invites us to consider in her poetic text "An Open Book of Grass": "Were we to be a Karner Blue Butterfly we would innately know the forest-ness of the prairie, grasses towering overhead, enfolding us."[23] To begin to engage with the complex beauty of grasslands, to no longer be blind to them, we must give up our human centrality, even if at first only imaginatively, and lose ourselves in a field of alterity charged with sheer potential to reconfigure and transform how we think and who we are.

Post-structuralist philosophers Gilles Deleuze and Félix Guattari found in the rhizomatic networks of grass roots a critique of traditional hierarchical ways of Western thinking. They mapped the intricacies and potentialities of the rhizome to illustrate an all-encompassing connectivity where any point can be linked to any other without a predefined or privileged path. This contrasts with the traditional "tree of knowledge" model that assumes a root or base that can only allow us to envision vertical epistemologies and genealogies of thinking that branch out toward the divine. The rhizomatic model knows no origin or destination. It is a decentralized and decentralizing network of critical nodes that sustains diversity. Embracing plurality, the rhizomatic paradigm rejects binary oppositions, stating that meaning and knowledge are never permanent or absolute but are continually shifting, morphing, and unexpectedly developing, thriving in the possibility of new connections and configuration.[24]

It is in this sense that grasslands do not provide static models for us to easily appropriate or simply see ourselves reflected in. They instead invite us to merge and connect existing knowledges in order to form new ones; to slow down and experience new forms of being present; to expand our thinking and being formats but without dictating how. The generosity of grasslands and their incomparable ecological wealth invite us to attune ourselves, to learn how to look again, to rethink hierarchies of scale and the depths of time, to train our eyes, ears, nose, and skin to become part of rhizomatic blueprints, in time, with patience, and by carefully recalibrating and attuning our attention to the wholesomeness of vegetal otherness—they radically reeducate us.

2 YARDS

Midway upon the journey of our life
I found myself within a forest dark,
For the straightforward pathway had been lost.[1]

It is no coincidence that Dante Alighieri's soul-searching epic journey should begin in the deep darkness of the forest—the entrance of Hell. Dense, and treacherous, in Dante's *The Divine Comedy* the forest is terrifying and bitter: a place where the "sun is silent."[2] Only by carefully considering the relationship between light, plants, and space can we fully grasp the narrative meaning of his "*selva oscura.*" *The Divine Comedy* is a complex allegorical work permeated by strong anthropomorphic currents. The forest is fraught with danger and uncertainty. Wild beasts repeatedly block Dante's path: a leopard representing lust, a lion standing for pride, and lastly a she-wolf symbolizing greed.[3] Danger lurks amidst the old and contorted trunks and branches—markers of spiritual desolation, turmoil, and corruption. Dante's journey reflects his existential struggle to find the right ethical and moral paths, to escape from the darkness of sin and ignorance, and to move toward the light of knowledge: the redemption of divine love.

During the Renaissance and beyond, *The Divine Comedy* quickly became an incredibly potent reference point for the arts. From painting and literature to theatre, it carved the representational mold that crystallized the forest's negative symbolic essence. But Dante did not invent the idea of the evil forest from scratch. Witchcraft, crime, and sexual promiscuity: the forest had for centuries been a place of deviance. It is in the darkness of the forest that paganism survived despite Christianity's will to fully eradicate it. According to Greek and Roman mythology, Bacchus's followers gathered in the deep darkness of forests at night to indulge in orgiastic rituals away from the prying eyes of society.[4]

The unknown vulnerability, the sleep of reason, utter loss—the forest is at once a physical place dominated by plants and a metaphorical dimension of the human mind. It inexorably exposes our alienation from the rest of the earthlings; it painfully reminds us of our existentialist inadequacy and inability to be a part of it.

The Divine Comedy was strongly influenced by the humanist ideals that supported the cultural positivism of the Renaissance. Fictitious conceptions of human exceptionalism that through the revival of classical grandeurs centralized and isolated the human from the coevolutions that bind all life forms on this planet. Dante's forest here alludes to more than just the darkness that descends over a person who has lost faith. It is the symptom of a much larger malaise called *anthropocentrism*.

As the trees lessen and light begins to filter through the branches, Dante comes upon a castle surrounded by an open field a "*prato di fresca verdura*"—a lawn of fresh grass. This space, peaceful and open, albeit tinged by melancholia, is Elysium: a paradise for virtuous pagans, akin to the Elysian Fields of Greek mythology, that Dante situated in the outermost circle of Hell, known as Limbo.[5]

By contrast, this serene clearing emphasizes the moral darkness of the forest. Elysium, which is separated from Heaven by the mountain of Purgatory, serves as the ultimate resting place for the spirits of unbaptized intellectuals, philosophers, artists, and other figures of culture and valor who were born before Christ. The light of their ethically worthy spirits, reasoning, and talent could not be relegated to the darkness of reason. Dante's description of Elysium is parsimonious: he only mentions green grass. But in Omer's *Odyssey* we find more: "to the Elysian plain . . . where life is easiest for men. No snow is there, nor heavy storm, nor ever rain, but ever does Ocean send up blasts of the shrill-blowing West Wind that they may give cooling to men." [6]

For centuries, throughout the arts, the dichotomies "dark forest/sin" versus "bright lawn/redemption" have dominated representational tropes as well as the structure of our thinking. Castles, as well as towns and cities, were indeed surrounded by lawns grazed by cows, sheep, and goats. And in truth, forests were dangerous. Not only wild animals but thieves and brigands lurked in its darkness. In a medieval world in which the cultural elites lived within the perimeter of fortified cities, towns, and courts, venturing outside was truly risky. This stark contrast between the forest's apparent irrationality and the orderly, bright surroundings adjacent to the city walls served to further emphasize the nature/culture divide. It is no coincidence that the word "forest" should come from the Latin *forīs*, meaning "outside."

Precolonial grassy lawns were often more like meadows strewn with wildflowers—buttercups, daisies, dandelions, poppies, and other species. However, the desire for aesthetic homogeneity and uniformity didn't take long to bloom. One of the earliest popular manuals for managing the feudal estate, *The Advantages of Country Living* by Petrus de' Crescentius, published in 1306, featured a detailed description of how to grow the most even lawn. To eliminate weeds, the author recommended first preparing the soil by pulling out all plants and roots before scorching it with hot water. A uniform turf, harvested from the wild, was then placed and hammered with mallets until all grass blades disappear. De' Crescentius guaranteed that fresh and even new growth would soon appear and that cutting the grass twice a year would suffice to keep it in shape.[7]

Lawns surrounding towns, castles, and monasteries were buffer zones—a domesticated quasi-nature, closely monitored and regularly repressed, welcoming to us but unhomely to wildlife. The aesthetic tidiness, the evenness—the lawn quickly captured the imagination of Renaissance gardeners across Europe. As far as composition is concerned, grassy lawns served as the necessary negative space between geometrically designed flower beds and to highlight architectural elements such as temples and fountains.

Careful selection of specific species of grass quickly gained importance too, as did the perfecting of meticulous shoring with sickles. "Ryegrass" (*Lolium*) was favored for its hardiness and ability to thrive in various soil types. Low maintenance and drought resistance, "fescue" (*Festuca*) and "bentgrass" (*Agrostis*) were appreciated for their fine texture and dense, carpet-like appearance. Paths covered in bentgrass carved their way through flower beds and groves, leading visitors on a journey along the various design elements and to picturesque viewpoints. Different grass types also expanded the textural and chromatic pallets of garden designers, eventually promoting the emergence of lawn design as an art form in its own right. Gradually, as garden design evolved, lawns were implemented in the parterre, a formal garden construction on a level surface, often composed of plant beds, edged in stone, and gravel paths.

In England and France, the large expanses of well-tended lawn gained popularity among the very wealthy during the seventeenth century. These greens became important for recreational activities such as bowling or simply wandering and admiring the scenery. As an open space to see and be seen in, the lawn quickly turned into a highly charged field of spectacles—a stage upon which the aristocracy could display their wealth and sophistication. The royal celebrations at Versailles, conducted between 1664 and 1674 under the reign of Louis XIV, propelled the lawn to international fame. No royal festival was complete without a ballet. Laborious and expensive, ballets would take months to plan and required extensive rehearsals. The king's participation was essential—his dancing virtuosity, a manifestation of his social superiority.[8] His body's synchronized and regulated movements demonstrated humanity's capacity to sublimate desires and untamed natural drives. Domination, control, and subordination: the lawn, upon which the dance took place, reflected this power matrix of social supremacy. An undisputed sign of regal distinction, the lawn's appeal quickly conquered the higher aristocratic strata as one of the most exclusive status symbols of its time.

More than other possessions, ostentatious and pristine, the tidiness of one's lawn indicated the current state of one's wealth. The more complex and ornate the garden design, the more expensive and demanding its constant upkeep: the perfect symmetry of topiaries and linearity of edges signalled the owner's opulence in effortless ways—at a glance, for everyone to see from afar. In a world where most people cultivated other people's land and those who owned land carefully maximized the potential of agricultural productivity, the aristocracy could afford to surround their mansions with sterile, unproductive, and extremely expensive expanses of green grass that required constant maintenance. Labor for the sake of aesthetics and virtue signalling—a luxury only a few could afford.

By the beginning of the eighteenth century, the aristocratic lure of the lawn quickly spread among the new mercantile class that vastly benefited from colonialist trades. Not only did they desire larger and smoother lawns, but they also wanted them portrayed in expensive paintings. Representational cross-pollinations in the construction of the landscape of empire: at this point the history of garden design and landscape painting became intimately interwoven, one influencing the other in complex aesthetic dialectics.

The transformation of the land, and its representation, during the seventeenth and eighteenth centuries was part of a colonial project of identity building in which the West promoted cultural, technological, and financial supremacy over the rest of the world. Paintings immortalizing mansions and castles followed a precise aesthetic logic grounded in the careful use of light and perspective designed to elevate and aggrandize. The pleasant and peaceful appearance of painted immaculate lawns aptly concealed the chains of exploitation, death, and ecological devastation that fed the grass' roots.

Aesthetically, these paintings belonged to the tradition that philosopher Edmund Burke called "the beautiful": an all-encompassing, imperturbable sense of harmony, balance, and serenity.[9] From Canaletto, Constant Bourgeoise, and James Gibbs to John Constable and William Turner, in these paintings, the lawn incarnated the dream of a highly exclusive lifestyle. Today, these canvasses do not evoke the same sense of majesty and seduction in us since the lawn has taken over the world. But back then, those works of art were instrumental in its popularization across empires. A florid trade of prints immortalizing the allure of European estates, mansions, and castles fed an unstoppable craving for lush lawns in the United States, South America, and Australia. Settlers constantly referred to European societal standards as a benchmark of civility in new homelands they thought filled with savages. In colonized territories, the lawn became the epitome of the ecological and cultural erasures that settlers habitually perpetrated in the rewriting of Indigenous histories and traditions. In North America, the Calvinists, influenced by European Puritan notions, closely associated untamed nature and its inhabitants with sin. From the state of New York to Pennsylvania and beyond, they thus proceeded to destroy much of the native vegetation and the fauna that depended upon it. The lawn became the actual and metaphorical blank slate upon which settler mentalities could construct their new identities, unimpeded, unhindered, and in plain sight.

Throughout the nineteenth century, as the industrial revolution relentlessly and drastically redefined urban as well as rural realities, technological innovations heralded a mechanized standardization of lawn maintenance. Animal grazing, scything, and horse rolling were swiftly relegated to the past—as the growing power and affluence of the new bourgeoise grew, so more efficient tools became necessary to speed up and standardize lawn care. In 1830, Edwin Budding, an engineer in the textile trade, invented the cylinder, push lawnmower. Budding's patent specification stated, "Country gentlemen may find in using my machine themselves an amusing, useful and healthy exercise."[10] His few remarks changed the course of global history, laying the foundations for a patriarchal mold that would later give rise to a multibillion-dollar industry. It is this specific mass-cultural psychological dimension that transformed a simple process of land maintenance into a deeply meaningful reinvention of masculinity for the twentieth century.

Steam-powered lawnmowers appeared in the 1890s, but by 1902, the manufacturer Ransomes produced the first commercial lawn mower powered by an internal combustion gas engine. By 1920, British Atco launched smaller machines that could be used in home yards.[11] As the sanitization of living conditions became more urgent in Western culture, the obsession with having a perfect lawn rose and became even more widespread across social strata. The aftermath of the First World War led to a heightened awareness of public health and hygiene. The war had mercilessly exposed the catastrophic effects of disease and inadequate sanitation on a broad scale. At that point, the rise of modern medicine and the establishment of health organizations also played a crucial role in reshaping our cultural framework, as did the discovery of antibiotics and vaccines in the 1920s and 1930s, which revolutionized healthcare and led to a greater understanding of the importance of cleanliness and hygiene in preventing diseases.

From architecture to institutional organization, and domestic spaces, the West was pervaded by a technologically driven positivism that promoted the rationalization and efficient capitalization of space through modular approaches. Mowing the lawn perfectly fits into this pragmatic, growing sense of ethical responsibility toward the self and others. A modern mode of societal care, a process of redemption enabled by technological advancements. Modern life's systematized rhythms and new role models overwhelmed masculine heroism, which needed new ground to stand on and prove its worth. Not quite the pioneer of the far west nor one of the many kingly Louises of France, yet in charge of mowing the front lawn, the office man and factory worker alike could carry out their patriarchal duties: curbing nature's unruliness to provide a safe haven for the family. From father to son, generation to generation, mowing grass thus became a modern, mundane ritual designed to mark, if only performatively, the new contour of a just and rational masculine domain in which colonialist, Christian, progressive, and sanitary values appeared seamlessly—and naturally—rolled into one.

The lawnmower reinvigorated the modern man—the keeper of the "good and proper"—boosting his masculinity through a performative reconnection with an already domesticated and tame nature that he could successfully maintain with relative ease. Grass mowing seamlessly reassessed a truthfulness: the membership to an exclusive masculine club. A clan of men that despite living in comfort still enjoys getting their hands dirty (just enough) with grease and mud. The outdoor physical action, dedication, learning of tricks and styles, the attention to detail. Author Tom Fort, in the introduction of his riveting book *The Grass Is Greener*, carefully and accurately describes grass mowing in its distinctive olfactory and tactile dimensions: qualities that transcend the material world to craft mythical memories of past patriarchal glories.[12] Melancholia and nostalgia: a ritual that unites all suburban men, an unspoken crucible in which taking care of grass surrounding one's home connects history to the present. On a much smaller scale, grass mowing performatively reenacts settler deforestation; it reassures that sin, which often comes in the form of spiky and hairy weeds, will be unforgivingly eradicated. Thus, the humble modern man swiftly ascends to the ranks of local-savior—his mower returning heavenly peace to the corruptness of a wild world.

As a mundane occurrence, this quasi-sacred, cathartic ritual endlessly repeats itself in the yard's micro-milieu. The resurrection of masculinity amidst the industrial revolution's rapid transformation is performed weekly throughout the summer heat. A matter of personal pride, a way to reconnect with a long and ancient genealogy of outdoorsy conceptions of manliness, courage, and prowess from which women have been almost categorically excluded.

At the beginning of the last century, the lawn emerged as a psychological territory in which physical strength corroborates the idea of mental health. Stereotypes: men play with machinery, it's natural to them; it follows that it is natural to them to play with machines in a "natural" setting. They get dirty, sweat, push, and pull. They are active. Men can and should have a rough side. The grass quickly becomes turf. One's exclusive domain: "every man's house is his castle," a castle surrounded by a lush lawn.

For decades, during the last century, anything that happened in the flowerbed was considered female domain—pretty and delicate color combinations—anything pertaining to grass was male competence and concern. In opposition to the frivolous whimsy of the flower bed, the lawn was a matter of continuity and integrity that only a motorized, and potentially dangerous, machine could maintain if competently used by a man.

At the center of an unlikely process of masculine redemption, saving man from the dullification of a progressively alienating world that insisted on separating him from nature, more essential tools came into being. Trimmers, rollers, watering hoses, sprinklers, and of course the ubiquitous leaf blower. Closely related to motorbikes and sports cars, much of the allure of mowers and blowers rests (in the minds of men) on the rumble. Over the past century, in the Western world, men have been culturally programmed to associate the rumble of engines with the prowess of masculinity—the deeper, the better. Like a baritone voice amplified to deafening loudness. The thundering sound of a mower is a man's primordial roar. Generation after generation this association has stratified into a cultural predisposition that often seems innate. In an urban or suburban setting, where homes are close to one other, loud mowers and blowers can drive neighbors insane.

In *American Green*, Ted Steinberg provides a detailed and vivid account of the legal wars and neighborly disputes that since the early 1980s have hit the news.[13] The hate, the feuds, the simmering vendettas. The strident noise of gas-powered motors is more than a nuisance and in the hands of overzealous homeowners, it can go on for hours, multiple times a week. Although it is common knowledge that excessive mowing can destroy a lawn, the detrimental effects that blowers have on gardens, land, animals, and people appear to be still overlooked by many.

Noise pollution emitted by gas-operated leaf blowers is a serious matter. These machines can produce noise levels up to 75 decibels at 50 feet away, which is well above what's considered safe for sustained exposure. This level of noise pollution disrupts wildlife, contributes to hearing loss in humans, and negatively impacts the quality of indoor and outdoor living for entire communities.[14]

Gas-operated leaf blowers also pose a significant environmental hazard due to their high emissions. They typically run on comparatively less efficient two-stroke engines. A substantial portion of unburned fuel is emitted directly into the air as pollutants. These include volatile organic compounds and nitrogen oxides that contribute to the formation of ground-level ozone. A 2011 study revealed that running a leaf blower for 30 minutes creates more emissions than driving a F-150 pickup truck 3,800 miles.[15] The powerful air jet is comparable to the devastating force of a wind blowing at 180 to 200 miles per hour: this can damage plants and kill insects and other creatures that dwell close to the ground.[16] More than any other gardening equipment, the leaf blower embodies the utter absurdity of our convoluted relationship with grass.

Mowers, blowers, and the pesticides needed to keep a lawn looking healthy, essentially are the antithesis of respectful neighborly living. We constantly poison each other and make our lives miserable for the sake of appearances. Neatly and regularly trimmed, the lawn in one's yard tells everyone in the neighborhood that dad cares. It is a matter of pride. A healthy-looking lawn casts the idea of a healthy and happy family: the kids, the wife, the dog, and the flowers on the windowsill. The lawn's normative power is enormous. It is no coincidence that in the United States, suburbia emerged in the 1950s as the physical manifestation of the American Dream, incarnated in the hellish-idyllic modularity of non-urban living wrapped by a white picket fence. Cookie cutter homes, garages, a sprinkle of trees, the ubiquitous cul-de-sac, and lawns: an unbroken connective tissue holding together the topographies of a far-too-quickly expanding, unimaginative urbanization.

Kids play. The football games. The barbeques. The cold beer with the mates. It's important that the lawn is kept clean, even, and well-tended—always. First and foremost, the lawn is an extension of the indoor spaces—a living rug or carpet. Its kempt state reflects the tidiness of the house's interior. It's a matter of *decorum:* the implicitly agreed upon cultural standard of good taste and propriety that all members of a community are expected to comply with. From this vantage point, it is evident that the front lawn is determined by distinct sociopolitical power dynamics that elude the backyard. The front lawn is at once private and public; it's the signature of a social agreement. Housing associations, municipal guidelines, and your neighbors have their eyes keenly set on it too. Missing regular mowing, in some jurisdictions, can land fines or even court summons. An empty space filled with responsibilities, burdened with the impossible task of ventriloquizing civic pride. The front lawn is not meant to be an honest and true reflection of who you are, it shouldn't portray one's vices and obsessions—it is a mask. It is emblematic of the veiled hypocrisy we all subscribe to for the sake of quiet communal living.

David Lynch masterfully unveiled this reality in the iconic, neo-surrealist opening sequence of his 1986 film, *Blue Velvet.* The idyllic scenes of bright flowers against picket fences and kids returning from schools to their suburban homes are swiftly interrupted as the family man (while his wife watches TV in the living room) suddenly collapses to the ground while watering the lawn. As the camera zooms closer and closer to the emerald blades of grass, a simmering world of worms and beetles festering just beneath the surface starts to show. It mercilessly points at the corruption and rot that lurks underneath the American Dream.[17]

Since the 1950s, the front lawn has been a moderate, virtue-signalling portrait that tells your neighbors just what they need to know: that you fit in, that as far as anyone should be concerned, your values mostly align with theirs; that you are no threat to their serenity. From a formal standpoint, the front lawn invites convivial, neighborly trust. Its regular maintenance and tidiness tell neighbors that everything is fine, things are under control, and you're on top of it. The front lawn is reassuring. It softly speaks of a stability that only patriarchy can maintain. A constancy and rigor that, mow after mow, shape character and civic rectitude.

The recent hyperpoliticization of the front lawn has further evidenced the remit of its socio-cultural power. Half a century ago it was the silence and emptiness of the front lawn that expressed virtue, but today a plethora of political signs has turned the front lawn into an extension of social media platforms. Yard politics war escalated in 2020 as the world faced unprecedented socio-communicational challenges brought by the Covid-19 pandemic, the rise of social justice, and the hotly debated Biden/Trump election. So much to say! Suddenly, the front lawn began to speak of what perhaps it never should. A collective cultural phenomenon grounded in the grass; a loud and clear message; a bold statement. No longer a visual equalizer, the lawn began to separate and polarize—openly, uncompromisingly.

Nothing to do with nature.

3 PITCHES

The Council on Foreign Relations' Global Conflict Tracker estimates that, at any given time, approximately thirty ongoing conflicts are raging worldwide.[1] As many as 828 million people—about 10 percent of the global population—regularly go to bed hungry.[2] Eighty thousand acres of tropical rainforest are lost to agricultural practices and fires every day. Scientists predict that climate change will cause more than a third of the Earth's animal and plant species to go extinct by 2050.[3] And according to a 2018 IPCC report, we have less than a decade to turn the tide on climate-induced catastrophes.[4] Meanwhile, social injustice is rife, right-wing governments threaten democracy, and neoliberalism is strangling cultural institutions around the world turning some into shopping malls and bulldozing others.[5] Given the current impending doom and gloom, it should not come as a surprise that so many people were outright upset by the continually bad conditions of grass at the Arizona Stadium as it prepared to host Super Bowl 57.[6]

"Atrocious," "deplorable," and "embarrassing"—these were only some of the strong adjectives used on social media to convey the unbearable disappointment.[7] And in a sense, there's reason for that, at least among the fans. The Super Bowl turf preparation takes two years. A team of experts incessantly monitors its growth, uniformity, and density. In this case, to assure the best possible quality, more than 600 rolls of turf—each 40 feet in length—traveled to the stadium from a local sod farm four weeks prior to the event. At Arizona Stadium, the turf slid over a giant tray that can be regularly rolled out of the venue to soak up the sun it needs to grow. The total cost was $800,000.[8]

The turf used was called Tahoma 31, a mix of Bermuda grasses (*Cynodon dactylon*) and Rye grass (*Lolium perenne*) developed at Oklahoma State University under the supervision of Dr. Yanqi Wu. Professional sport standard turf must perform extremely well under intense pressure, trampling, and far from ideal climatic circumstances. Stadia have over the past century evolved from open-air arenas into cave-like hovels. Sheltering the seating areas from rain to justify, at least in part, the exorbitant ticket prices has come at a cost. The rafters cast shadows; the turf wilts.[9]

The State Farm Stadium turf let the players and fans down. It was criticized for its unevenness. This inconsistency in the playing surface was not only a tripping hazard but also affected the roll and bounce of the ball, thereby impacting the overall quality of the gameplay. Despite being a grassy surface, the field was as hard as concrete in some areas— something that increased the risk of severe injuries to players. The grass lacked resilience and could not withstand the high-intensity play typical of a professional football match, leading to significant wear and tear even in the early stages of the game. The pitch appeared patchy and discolored.[10]

Does any of this really matter? When did sport stop being a game that simply brings people together?

The Super Bowl is, first and foremost, a money-making machine. Advertising revenue can generate around $550 million.[11] Hosting the game can bring in more than $1 billion to local economies.[12] The ticket sales alone can gross more than $65 million.[13] The various NFL partnerships involved in the big game have in the past generated as much as $1.8 billion.[14] Despite the occasional patchiness, sport grass is undoubtedly among the greenest in the world, in many ways.

But just as a yard lawn is never just grass, so the pitch turf is a complex field of power relations, intensities, genealogies, and potentialities. Its livingness, health, and aesthetic appeal matter a great deal. Among the blades of grass are rooted ideals of authenticity, tradition, and patriarchal dominion that play crucial roles in the game's philosophical integrity, the identity constructs that surround it, and the intensity of the feelings experienced by fans. Sports turf is emotionally charged like no other kind.

One might wonder why, if sports turf is so hard and expensive to maintain, not switch to artificial alternatives? In truth, this shift has happened in some places. Artificial lawns need no watering, mowing, fertilization, and pesticides, which can save significant time and resources. At a stretch, despite their plastic nature, artificial turfs might be considered more eco-friendly. And they most definitely provide a consistent playing surface, which can be particularly beneficial for sports where the condition of the field significantly impacts the game.[15]

But safety and eco-friendliness seem to matter less when a prestigious event is involved—tradition and authenticity take precedence. Outdoor sports like rugby, cricket, baseball, and football—those played on grass turf—belong to a different class, at least from historical and philosophical perspectives.

Team sports are miniature simulations of imaginary or latent conflicts. Mini-armies battle each other to the bitter end. It is no coincidence that the language still used in the realm of American football should be drawn straight from military combat. Terms such as "blitz," "bomb," "bombardier," "neutral zone," "red zone," "trenches," and "field general" reflect the strategic and confrontational nature of these games.[16] During the apartheid years in South Africa, rugby served as a metaphor for the country's racial and political struggles. The Springboks, South Africa's national rugby team, became a symbol of white supremacy. The 1995 Rugby World Cup, hosted by South Africa, turned into a metaphor of the country's struggle and eventual triumph over apartheid, as represented in the movie *Invictus*.[17]

The infamous "Soccer War" between Honduras and El Salvador in 1969 was triggered by tensions during a match. Although the root causes of the conflict originate in deeper socioeconomic issues, a soccer match served as a catalyst and a metaphor for the larger conflict.[18] And during the First World War, the "Christmas Truce" of 1914 saw British and German soldiers temporarily cease hostilities to play a game of soccer in no-man's-land. This event has been widely interpreted as a metaphor for the human spirit and commonality that exists even amidst the horrors of war.[19]

Still sport/war entwinements are much more ancient. Cuju is one of the oldest forms of outdoor sport played by opposing teams over a grassy field. It started in China, during the third century BC as a fitness training for military cavaliers. As it gained traction among the nobility and imperial courts during the Han period (206 BC–AD 220), it evolved into a popular pastime, quickly spreading to Japan, Vietnam, and Korea. The game was played on a rectangular field (half the size of today's football pitches), with a goal in the center, marked by a rope or low walls. Cuju might be a rightful ancestor of modern football.[20]

Plato and Socrates often used athletic competition as a metaphor for warfare. They saw sports fields as complex spaces charged up by invisible power relations that allowed individuals to demonstrate virtues like courage, discipline, and strategic thinking. Just as athletes strove to win within the rules of their sport as outlined by the field, so too warriors aspired to win within the rules of just warfare. The field is more than a space for play. It is laden with rules and conventions one must learn and respect. It is a disciplinary spatialization. It forges the players' character and by extension that of the audience.

In Greek and Roman times, more similar to rugby, Episkyros and Harpastum were played in rectangular grassy fields especially in preparation for battles. Involving a strenuous regime including running and physical contact, these games were part of an essential training and mental conditioning routine. They carved the physical strength, agility, strategic thinking, and competitive spirit of the players, preparing them for the challenges of war and life.[21]

In the Middle Ages, this metaphor was further strengthened by philosophers like Thomas Aquinas, who saw war as a kind of "moral sport." Aquinas argued that just as in sports, where rules are established to ensure fair competition, so too should there be rules in warfare to prevent unnecessary suffering and destruction.[22] These considerations may have stemmed from observation of the often unruly and violent sports played at this time. Medieval sports could get rough, swiftly. Jean-Paul Sartre and Jacques Derrida saw sport as a form of "controlled war," where the same competitive spirit and tactical thinking apply. Herbert Spencer also, controversially, suggested that both sport and war are driven by the same basic human instinct for survival and dominance, with sport being a more civilized expression of this instinct.[23]

Cricket and other bat-and-ball games, like rounders, were played on very uneven and patchy village greens and fields. Still today, the contemporary oval shape of a cricket field is a nod to the humble sport's origins in rural England where it was played on irregular grounds—a reminder of the sport's philosophy of adaptability and resourcefulness. Grassy pitches were often loosely marked with goals far apart at times miles away from each other, as it was the case with Shrovetide football. Believed to have originated in the twelfth century, the game was traditionally played on Shrove Tuesday and Ash Wednesday, the days leading up to the Christian season of Lent. The name "Shrovetide" comes from "shrive," which means to confess sins, a common practice during this period. The game was the last chance to indulge in merriment and revelry before the solemn period of Lent. Chaotic and often violent, Shrovetide football involved entire villages. Kicking and chasing an inflated pig's bladder, players aimed to score a goal by any means necessary. Associated with ancient fertility rites and pagan traditions, the game was believed to bring good fortune and abundant crops. Despite attempts by the authorities to ban it due to its violent nature, it persisted for centuries, evolving over time.[24]

The institutionalization of sports that took place between the seventeenth and nineteenth centuries drastically consolidated and strengthened rules and regulations, further limiting bodily interactions and systematically repressing spontaneous violent bouts. The layering of regulations that allowed and prohibited behaviors in outdoor team sports carries a quasi-performative character that mirrors societal structures and subtly promotes the notion of the survival of the fittest.

Lines, zones, and areas hierarchies and subordinations. Rules of conduct and range of actions are dictated by parameters and intersections; white lines on green grass delimit borders—inside and outside—and divide property. A distilled plateau of social control, a closed system of polarized and endless variables.

Both incarnations of European and American football emerged during the nineteenth century—a time of cultural and civic rationalization and standardization—as the industrial revolution relentlessly reconfigured everyday life, radically remodeling the relationship between countryside and urban realities as well as generating a sense of further alienation from the natural world.

American football originated at universities in New Jersey, and it is the descendant of English-born soccer and rugby. On November 6, 1869, Princeton and Rutgers players competed in the first intercollegiate football game in New Brunswick.[25] Harvard and Yale followed suit in 1875. Legend has it that in 1823, while attending a private school in England, a young man by the name of William Webb Ellis inadvertently invented rugby when he raced to the goal line of the opposite team, holding the ball in his hands. It is telling that the modern iterations of both games, which require the maintenance of expensive grass lawns, emerged from institutional elitist Ivy League universities or private schools. In 1827, Oxford and Cambridge institutionalized cricket, making it the first university athletic event in the UK to be documented.[26]

Field sports became a central part of university life for several reasons. First, like strenuous training required by war, sports built character and instilled values such as teamwork, discipline, and perseverance—especially in men. Simulations of future professional interactions and open-air team sports like rugby and football prepared male students for leadership roles within an institutional structure that sidelined women as fragile, unreliable, and inept. It was upon the green turfs of university sports grounds that ancient battle tropes from China, Greece, and Rome were refashioned into highly competitive legal and business training fields. It is no coincidence that both professions also incorporate war expressions like "minefield," "take the flak" (firing a bullet at an aircraft from the ground), "lay low," "lose ground," and "making a killing" in their jargon. Battling one another, teams would also metaphorically "go to war"—an army defending their universities' reputations—an extremely lucrative marketing opportunity for educational institutions across the United States.

This extensive genealogy that intimately entwines field sports and war requires living grass to retain its essential, ideological integrity. A symbolic token of truth, the living grass turf becomes a transcendental plain that, much more powerfully than even the yard lawn, conjures millennia of societal structure building, antagonism, and systems of symbolic confrontation. In outdoor sports like football and rugby, the bouncing of the ball back and forth across the field represents the constant gains and setbacks of war battles.

Groomed and tamed, the pitch is a material abstraction; the living memory of the land upon which the conflict unfolded many centuries ago. In war, the winner almost always claims fertile or strategically important land. The livingness of the grass underpins the veracity of the simulation; it lends conceptual credibility to the performance by symbolically linking it to a natural order of dominance that humans must uphold in order to stay alive. The sports pitch is therefore a transhistorical plateau across which the primordial fight between good and evil relentlessly unravels and sinister fantasies of usurpation can be performatively fulfilled.

Some have argued that competitive team sports are an essential part of human social and cultural predisposition and that they express and allow the peaceful manifestation of natural desires. However, on the other hand, their conflict-grounded ideological matrix simply perpetuates negative values that shape, from a very young age, the minds of millions into simple binary dimensionalities that ultimately legitimize rivalry and promote hate.[27]

To simply focus on turf sports as war-miniatures, oversimplifies the complexities and histories embedded in very special kinds of lawns. Slovenian philosopher Slavoj Žižek sees sports as a social phenomenon that by far exceeds war-simulation. In his views, sports more deeply reflect the underlying ideologies and power structures of society where the rules of the game, with their pre-inscribed and infringeable roles and functions, mirror the rules of normative and disciplined social conduct. Violations, such as fouls or cheating, are penalized, reinforcing civic moral dimensions. Furthermore, sports rules are not up for discussion or negotiation, they are part of an authoritarian and rigid administrative regime that predisposes and primes large crowds to unquestionably accept and follow the absolute power of tyrannical governmental forces into ideological submission.

In the rules and ethical dimensions of field sports also lie a reflection of hypocritical superficiality. The conception of "fair play" entails more than following the rules. It maintains the illusion of a level playing field in which all players equally matter. This implied configuration, Žižek suggests, is akin to how capitalist societies promote inherent equal opportunity ideas while ignoring underlying structural inequalities.[28]

The gender and social normativity that competitive team sports may performatively propagate and perpetuate are also visible in the strong enthusiasm and fanaticism that supports and validates social identities and pure nationalism in simplistic and commercialized ways. Professional sports have become manifestations of capitalist exploitation where athletes are fetishized as commodities by a purely profit driven industry. This perspective aligns with Žižek's broader critique of capitalism's dehumanizing effects—a process that, since the end of the Second World War, has instilled a dangerous shift from substance to spectacle. The medium becomes the message. The masses are drawn to the drama of the action rather than to deeply engage with actual content or substance.

It is in this context that the green turf of field sports emerges as a site of modern normative validation. Social norms can be rigid, unchanged, and unchanging. The sports pitch is permeated by a steep conservativism that comforts and unites in inflexible certainty—one inscribed in the symbolic natural order evoked by the living grass. Repetitive and pre-encoded experiences clearly either allowed or forbidden, where only certain modes of bodily contact and conduct can exist and in which competition, strength, and winner/loser or strong/weak dichotomies are the only possible outcomes of human vicissitudes: field sports often leave no room for empathy and diversity. Working in subliminal ways, the symbolic and metaphorical dimensions of the pitch are unproblematically absorbed by the masses; weekend after weekend—in the name of fun and recreation—they shape behavioral predispositions by relentlessly shrinking thinking paths.

In *The Structure of Behavior*, Maurice Merleau-Ponty clearly explains how the process works:

> For the player in action the football field is not an "object," that is, the ideal term which can give rise to a multiplicity of perspectival views and remain equivalent under its apparent transformations. It is pervaded with lines of force (the "yard lines"; those which demarcate the penalty area) and articulated in sectors (for example, the "openings" between the adversaries) which call for a certain mode of action and which initiate and guide the

action as if the player were unaware of it. The field itself is not given to him, but present as the immanent term of his practical intentions; the player becomes one with it and feels the direction of the goal, for example just as immediately as the vertical and horizontal planes of his own body.[29]

This subtle and all-pervasive embodiment also stretches in its own form to the spectators that erupt in jubilation or utter consternation in reaction to actions performed by billionaire footballers—actions that will bear no positive impact on the fans' economic, familial, or employment realities. Competitive team sports thus reinforce a capitalist ideology where success is measured solely by individual achievement and profit. The tears, the laughter, the screaming, the racist and homophobic slurs—the pack mentality that enables and validates all of it: this is how humanity comes together to alienate itself. Like the grass that grows on the pitch, this deep emotional investment is real and yet wholly misplaced. The normativity embedded in the lines traced on the green grass, the "me" versus "you" structure inscribed in the field-zones, symbolically validates the naturalness of these behaviors too. Is it a coincidence that homophobic and transphobic abuse should be more frequent in football and rugby than in other sports?[30] In these sports, the green grassy field is materially confined to the perimeter of the playable area and yet it extends in all directions into the players' bodies and minds, reaching deep into the fibers of the spectators who passively identify with players and teams. This concept, thus, underscores the embodied nature of sporting activities and the role the grassy field plays in shaping these communal experiences into a monolithic, naturalized cultural identity. The crowd becomes an indistinguishable, homogeneous multitude, a sterile monoculture, just like the blades of grass on the pitch: only existing to rake in money for already exceedingly wealthy sports club owners.

Nothing to do with nature.

4 PARKS

Laughter and chatter—a group of women walk along the path—skinny jeans and branded T-shirts. They just ate together under the trees' shade. Nearby, a mother rides her bike, her daughter in tow. A red ball is tossed into the air. Girls play volleyball without a net. Winning and keeping score is not the point. Enjoying a few hours of freedom in the open is all that matters. The grass is green; the lawn is well-kept. In Tehran where traditionally women enjoy less social freedom than men, a women-only park represents more than fresh air and recreation. Mother's Paradise Park provides the opportunity to express one's identity beyond the stronghold of traditional customs. This is one of five such parks in the city. Similar spaces exist in India, Saudi Arabia, and Pakistan.

This park has over time become a symbol of liberation where women can express themselves freely, both physically and emotionally, away from the otherwise omnipresent gaze of patriarchy. The green spaces are utilized in various ways by women of all ages. For some, it's a place for jogging, yoga, or aerobics, which they can practice without the constraints of a hijab. For others, it's an opportunity to interact with friends, have picnics, relax, and participate in group activities, or a safe space for mothers to play with their children.[1]

But despite its popularity, Mother's Paradise Park is not without controversy. Some conservative factions in Iran view it as a threat to traditional norms and values. They argue that the park promotes Western ideals and undermines Islamic principles. In truth, this is a place where social issues, women's rights, and other topics that might be considered taboo in more public, mixed-gender spaces can be discussed. In this way, Mother's Paradise Park is also a place of resistance—a vital social and political space for Iranian women. In the eyes of many, the park has become a catalyst for greater gender equality. Many women see it as a significant stride toward a more equitable society.

However, the power structures that hold together the park's social freedom and constraints have been entirely designed and built by men. Iron walls and guardians: Are women-only parks spaces where patriarchy has structurally manifested its potency with impeccable precision? Has gender segregation reached its obvious conclusion? Why were female urban planners and landscape designers sidelined from the development of women-only parks?[2]

Referencing Elizabeth Grosz's influential essay "Bodies-Cities," sociologist Reza Arjmand has posited that "the politics of power are always sexual, even though space is a central mechanism of the erasure of sexuality."[3] In that sense, urban spaces are never neutral, they are fields of intensity permeated by ideological forces that actively, and yet invisibly, model our behavior. Visibility can be normative. It can hold in place; it can prevent and deter; it encodes. It identifies and hierarchizes. At times, it exposes to shame, but it also offers shelter from violence.

Gender is not a concept naturally inscribed in the body, but it is a nuanced conceptualization that emerges from the performativity of the body in everyday life. According to Judith Butler, the way we move in space, where we are allowed to go or not—where we should be seen doing what and speaking to whom—the apparent freedom of outdoor spaces is highly deceptive.[4]

It is in this context that lawns can be taught of as all-important optical devices—open and unencumbered swaths of green land: fields of absolute visibility. Lawns are implicitly panoptic in nature: spaces in which power relations constantly and subtly shift but where parameters, perimeters, and immaterial and yet unsurmountable barriers determine our identities and ways of being.

Panopticism, as theorized by Michel Foucault, is an actual as well as metaphorical, structure of power and surveillance. Far from being a theoretical abstraction, the panopticon is an actual prison model based on the architectural design by eighteenth-century philosopher—and father of utilitarianism—Jeremy Bentham, consisting of a circular building with an observation tower at its center, from which it is possible to observe all inmates without them knowing whether they are being watched. Foucault used the panopticon as a blueprint to illuminate the power of the gaze and its function in the disciplinary mechanisms of modern society.[5] He argued that the major effect of the panopticon is to induce a state of conscious and permanent visibility that assures the self-perpetuation of power structures. It is guard to be actually present in the central tower. The unverifiable possibility that they might be there, observing their every move, keeps the inmates in check. The field of visibility of the panopticon thus instils self-regulation through utter and incessant fear. Its mechanism operates in the prison of the mind. Efficient and cost-effective, it is continuously self-enforcing and, in time, it naturalizes itself as a seemingly neutral and unbiased social mode of conduct. A subtle and relentless self-disciplining that shapes individuals into compliant subjects who police their own behavior as well as that of others.

Panoptic forces are relentlessly at play in urban park planning. While a well-planted tiny park nestled among the austerity of city blocks might recall a picturesque oasis, trees and bushes can also literally get in the way of social panoptic policing. When this happens, some people feel deeply anxious or even threatened.

Just one week before the Stonewall Riots began, on June 18–20, 1969, a gang of local vigilantes cut down thirty trees in Kew Gardens - a small park in Queens, NYC Their objective was to remove visual obstacles to dissuade night-time, brazen meetings between gay men. It has been alleged that the vigilantes worked with the implicit support of local police forces. A few photographs documenting the aftermath show a desolate lawn dotted with tree stumps. The incident divided the local community. No arrests were made. In a matter of days, the gay rights group Mattachine Society began to raise funds to replace the trees.[6] The event was quickly overshadowed by the Stonewall Riots but it nonetheless sparked the earliest gay liberationist environmental movement.[7]

The deeply entwined patriarchal matrix of the lawn often results in a picture of desolation in which metaphorically any kind of vegetal natural life other than the reassuring and domesticated carpet of grass poses a threat to the supposed, clean-cut rectitude of masculinity. Normative, consistent, coherent, and most importantly straight, the lawn is uncompromising in its aesthetic integrity—a quasi-fascist bundle of alluring qualities rooted in the illusion of masculine grandeur. In the introduction of her seminal collection *Queer Ecologies: Sex, Nature, Politics, and Desire*, Catriona Sandilands shows how, in response to the relentless urbanization of life that marked the first few decades of the twentieth century,

> white men came to assert their increasingly heterosexual identities in the wilderness explicitly against the urban specter of the queer, the immigrant, and the communist: a legion of feminized men who were clearly not of the same manly calibre as the likes of Theodore Roosevelt.[8]

Mountain peaks, valleys, and streams—during the height of the industrial revolution, modern masculinity in the West emerged as a confrontation with the primordial and uncontainable forces of the natural world. This display of strength and tenacity was implicitly opposed to the effeminate fragility of women (and lesser men) who were, according to patriarchal views, better suited to the safety of the domestic sphere. In the absence of true wilderness, within the regulated bounds of the urban landscape, brandishing chainsaws to fell trees in and of itself constituted a performance in which masculinity was reclaimed through a confrontation with nature—or, more precisely, a pantomime through which gender normativity implanted in infancy was reaffirmed in adulthood. Chainsaws abated anxiety. The felling of trees in Queen's Kew Gardens was a reenactment of the ancient deforestations that kept wild beasts away from villages during the Middle Ages. Or even of those carried out by North American settlers, who hoped by so doing to cleanse the land from sin and perdition to moralize Indigenous populations. This event activated the power of the unencumbered lawn as a sanitizing agent—a process that simultaneously animalized queerness. The sunny airiness of the lawn, a quality always and inescapably drenched in Christian morals of divine rightfulness, is no place for the beasts of darkness. Tellingly, in the highly influential 1870 landscape design manual *The Art of Beautifying Suburban Home Grounds*, Frank J. Scott clearly spells out that:

> It is unchristian to hedge from the sight of others the beauties of nature which it has been our good fortune to create or secure and all the walls, high fences, hedge

screens and belts of trees and shrubbery which are used for that purpose only, are so many means by which we show how unchristian and unneighborly we can be.[9]

For this very reason, the openness of lawns has played a prominent role in urban parks. The domestic, the holy, the masculine, the sanitized—regardless of its specific geographic location, in the West, a lawn is deeply rooted in positivistic values of health, physical as well as mental.

Green spaces like Central Park in New York City were implicitly landscaped on these ideological premises. A green gem instilling morality right where it was more desperately needed, at the heart of (as some thought) one of the world's most polluted, corrupted, fast-growing, and chaotic cities in the world. The park, designed in 1858 by Frederick Law Olmsted—the first of its kind in the history of the United States—set the blueprint for the ones that followed. At that time, the swath of 758 acres designated by the local municipality to become Central Park was far from the virgin paradise Olmstead envisioned. The area had been deforested and built on already. While it was depicted in the local press as a rundown and insalubrious neighborhood, it is now recognized that it was the home of a vibrant Black and immigrant population that had settled there in 1825. For over thirty years, before it was forcefully seized by the city authorities in 1853, Seneca Village had occupied a stretch of land between 82nd and 89th Street.[10]

Central Park emerged from multiple erasures. A wholly artificial image of nature was then carefully assembled and manicured to create a timeless illusion that it had existed right there well before anything else, untouched, all along as if the city harmoniously grew, like grass, around it. But nature wasn't necessarily Olmsted's primary source of inspiration. His bucolic masterplan synthesized the classic design of European parks with paintings of the British countryside that in art historical terms belong to the genres known as the *pastoral* and the *picturesque*.[11] Canvas after canvas, between the sixteenth and eighteenth centuries, artists like Giorgione, Titian, Nicolas Poussin, and Thomas Gainsborough idealized the rural and natural world as a placid and benevolent space. Composed like theatrical sceneries, these landscapes embodied an inherent simplicity, balance, calm, goodness, and purity that can only reside in the spontaneity of nature and thus in the soul of the purest of men. These paintings reflected the philosophical ideology of the Arcadian ideal, a utopian concept derived from ancient Greek philosophy—particularly from the works of the Epicureans and Stoics—in which humans live in a state of incorruptible harmony with a tame natural world. Peaceful, idyllic rural scenes had an enormous impact on the history of landscape, as well as gardening and park design, across the West and have in time naturalized in the collective subconscious as an appropriately pleasing, essential conception of nature.[12]

Olmsted's Central Park is structured around an assertive and essential kind of contrast. Within its bounds, he carefully wove the warp and weft of a recursive narrative that can't be escaped. The park's topography, with its sublime glacial boulders, mounds, slopes, and sunken roads, was designed to optically deny its own artifice and erase its perimeter. Olmsted wanted us to continuously get lost—geographically and spiritually—again and again so as to experience the dramatic passages from darkness to light and bask in the joy that only open lawns can bring as one's eyes adjust to the sudden brightness to refocus.

In essence a domesticated Yosemite, Central Park takes the Sunday stroller through emotional winding trails: a miniaturized and gentile reenactment of that "confrontation with wild nature" that ultimately makes us human. The process repeats itself subliminally, stroll after stroll, as a performance we memorize. This emotional repetition, engrained in Central Park's every acre, trains us to appreciate the comfortable safety of the open space in contrast to the uncertainty cast by the overgrown vegetation. It leads us to instinctually, and therefore presumably naturally, seek the vast sun-kissed openness of the green grass meadows. Sight lines, perspectives, shifting scenes, shadows, and light—not only Olmstead's vision was truly spiritual, but it was also painterly in the sense that his lawns, like Sheep Meadow in the South End of the park or the Great Hill, on the West Side, were only meant to be admired and not walked on.

Like many of his contemporaries, Olmstead believed that lawns were central to an elevating and healing but purely aesthetic experience. Thus, the lawn was not for play or leisure, but first and foremost it was a picture, a living representation. As an open scopic field, its primary purpose was to corroborate the soul, not the body. Physically excluded from the lawn, only able to gaze at it from a distance, the park visitor contemplates a fictitious natural virginity. Colonial ideologies were Olmstead's frame that wrapped around every painterly scene he brought to life in Central Park.

It was only in the 1920s and 1930s, as modernist concepts of physical and mental wellness took precedence over romantic values, that people were allowed to lie and play on the grass. However, the pictorial essence of lawns in parks never truly subsided. In fact, it still survives today, deeply rooted in the very artificiality of its presumed naturalness.

In their homogeneity lawns imply an inescapable fragility. Nowhere the infamous "keep off the grass" signs have been more popular than in city parks, just where access to green spaces is so badly needed. By the 1970s, inadequate maintenance and heavy traffic turned Central Park's Great Lawn—a 15-acre field of green grass—into "The Great Dust Bowl."[13] In 1995, after a total of 200,000 people descended onto the Great Lawn to watch a screening of Disney's *Pocahontas* and attend mass by Pope John Paul II, the city decided to invest $18 million to bring the dilapidated lawn back to life.[14] A watering system was installed. Soil and sand were brought in from Long Island and half a million square feet of Kentucky blue grass was planted. It took two years to fix the lawn. The picture-perfect result was closed off to the public for months thereafter as New Yorkers waited for clear guidelines and regulations on traffic and usage to be issued.[15] For many years, the lawn became a political battleground as demonstrators were often denied permits to protect the grass.[16] In 2023, a third of the Great Lawn was destroyed once more as a concert went ahead while storm Ophelia relentlessly soaked the ground. Sixty thousand stomping feet reduced the grass to mush causing a six-months-long shutdown that reignited the never-ending perfect lawn controversy.

Nothing to do with nature.

5 GOLF COURSES

Dawn. Shards of glass all over the floor. Tipped over shelves and smashed florescent tubes; scattered soil and trashed rolls of turf. The noxious smell of spray paint still lingers in the air. A clue: golf balls emblazoned with the letter "A" dot the ground of the devastated greenhouses at Pure-Seed Testing research facilities, some miles south of Portland (Oregon). A few hours later, the act of vandalism is claimed by the Anarchist Golfing Association—a group whose stated mission is "to make life miserable for companies that develop genetically engineered grass for a sport they say is solely 'the pleasure of the rich.'"[1] A Pure-Seed Testing representative confirmed that the protestors "pulled and threw and stomped and mixed labels and threw away labels and did everything they could to destroy things."[2] This was only one in a string of crimes against similar facilities carried out in 2001 by biotech saboteurs who oppose genetic engineering. Pure-Seed Testing was specifically targeted because of their work on creeping bentgrass (*Agrostis stolonifera*), a species native to Eurasia and North Africa that is thought to have arrived in North America during the nineteenth century.

A highly versatile, perennial grass that thrives in cool, humid climates, initially, creeping bentgrass was employed in pastures and lawns due to its ability to form a dense, green carpet-like surface. However, it was its adaptability to different climates and soil types that made it perfect for golf courses. Its resilience to close mowing, creeping growth habit, and its uniformly fine texture proved major assets.

Over the years, various cultivar varieties of creeping bentgrass were selected and developed to enhance the species' most desirable traits: disease resistance, heat tolerance, and even finer textures. This continuous improvement has further solidified the position of creeping bentgrass as the favored species for golf courses around the world.

Its commercial success sprouted in the early 1900s, aafter groundbreaking research by Drs. Charles Piper, Russell Oakley, and Lyman Carrier, with support from the US Department of Agriculture (USDA) and USGA Green Section. The breeding programs that followed led to the engineering of several varieties including the Penncross, developed at Pennsylvania State University, which offered improved disease resistance, heat tolerance, and overall higher turf quality. Officially brought to the market in the 1950s, the grass quickly became the quality standard with golf course superintendents.[3]

Toward the end of the last century, the advent of genetic engineering led to the development of new, modified (GM) creeping bentgrass. The first GM bentgrass, "Roundup Ready," was developed by Monsanto and Scotts Miracle-Gro in the late 1990s.[4] Primary objective: to enhance weed control on golf courses through the development of grass resistant to the pesticide *glyphosate*. The engineering of "Roundup Ready" bentgrass marked a significant milestone in the history of GM crops. It highlighted the commercial potential of GM in the development of varieties with improved characteristics for golf courses, but it also clearly showed how ecologically damaging these manipulations can be.

In the summer of 2003, windstorms swept the Oregon GM testing fields, scattering bentgrass pollen and seeds well beyond the originally designated controlled growing area. Reports of herbicide-resistant bentgrass flourished far and wide for years thereafter. Farmers especially found it impossible to prevent the new super-grass breed from smothering crops and obstructing canals and ditches. All in the name of golfing . . .

The golf turf is among the most significant monocultural greens in world history. From an ideological perspective, its aesthetic perfection and smooth uniformity encapsulate every power structure that defines other kinds of lawn: the patriarchal values; the desire to subjugate nature; the drive to perpetuate normative models; the wish to sanitize the world; the spiritual elevation; the societal elitism; the moralizing undertones. But perhaps even more than just the sum of these ideological layers, the golf course turf is also a living emblem of supreme capitalist commodification. More than any other lawn, golf turf is in every sense *hyper*. The incarnation of a thoroughly "commodified pastoral," its vast expanse induces a kind of aesthetic experience characterized by deadly imperturbable serenity.

The pastoral style that emerged in Western painting between the sixteenth and eighteenth centuries in Europe proposed a romanticized view of the world where shepherds happily strolled around an idyllic countryside. Rolling hills and sun-kissed meadows provided the perfect stage for a morality play: an idealist rectitude—natural in appearance and yet wholly manufactured. While the pastoral genre could be viewed in a broader cultural sense as a response to the social and economic changes brought about by the industrial revolution and urbanization, its relentless idealization of the rurality and rusticness primarily aimed at erasing the harsh reality of agricultural and farming life in order to facilitate exploitation by the upper classes.

After all, it wasn't farmers or peasants who painted these pictures but artists who craved notoriety, financial gain, as well as the favors of the wealthy landowners who commissioned them. Although it depicted the countryside, the pastoral was an urban product; lush and bright canvases immortalized the vision of the landowner's gaze as an ultimately benevolent and magnanimous carer of both: land and laborers. These landscapes ennobled the gentry on both ethical and moral grounds by constructing a nostalgic vision of a lost world (that never was) of natural harmony and imaginarily harmonious social cohesion and order that only landowners, with their education, adequacy, and power, could truly support, nurture, and perpetuate. Arcadia—an idyllic and unproblematic rurality: the gentry masterminds and the peasants duly actualize their plan, silently, happily unquestioning—that's their lives' purpose. Their privilege: to work for them for next to nothing.

As a genre, the pastoral is more than conservative. It is regressive. Its oversimplification of reality is childlike, as argued by Friedrich Schiller at the very end of the eighteenth century.[5] The pastoral is redemptive, as long as we allow ourselves to stare at the painting long enough to lose ourselves in its lie.

Over centuries, pastoral ideals have sent deep roots beneath the smooth turf of golf courses: a commodified, hyper-manifestation of these invisible cultural forces. No cows nor sheep now roam the grassy slopes. The serene greenness of the pastoral genre, a perpetual promise of peace and plenty, has been exasperated to an artificial point of no return. Horses have turned into carts; shovels into clubs; shepherds into businessmen—like the shepherds populating the paintings of Jean-Honoré Fragonard, Abraham Cornelisz, Antoine Watteau, or John Constable, apparently wholly benevolent and well-meaning, they hang out and meet and greet each other in the fields.

Over the last century, new forms of pastoral capitalism have emerged across the globe. They are all rooted in the ideological values that permeate this longstanding artistic tradition, and they all aim, just like those paintings, to conceal and obscure the pain of others. As Louise Mozingo argues, "The landscapes of pastoral capitalism are part and parcel of corporate ideology, reinforcing the self-regard of the corporate management and mystifying the unpleasant actualities of corporate power."[6] Pastoral aesthetics, anchored in the lushness and serenity that only smooth and even lawns can convey at a glance, are subtle and powerfully persuasive. They deeply ingrain themselves in our minds: they turn into a symbol we instantly perceive, subliminally a symbol that has naturalized itself to the point that it no longer needs decoding.

In the case of golf courses, capitalist modularity has appropriated and repackaged the idealized simplicity of an imaginary rural existence. The fields are uniformly grazed by riding mowers; the grass is regularly watered and doused with up to nearly thirty pesticide-active ingredients designed to nihilate insect and foreign vegetal lives of all kinds. An average golf course normally receives twice as many pesticides as agricultural grassland.[7]

But ultimately, the golf course lawn, like all others, is primarily a picture; its aesthetic value is grounded in the apparent naturalization of a capitalist social order that keeps corporations in power. This subtle and yet all-pervasive representational operation hinges upon one key aspect that golf courses have unproblematically inherited from pastoral paintings: the ungrounded and yet confident promise of an uncomplicated life, detached from the toils and tribulations of everyday life. Escapism for the very rich, over and over again, endlessly, forever. Just as the paintings erased and enabled over 300 years ago, so do the golf courses today. Like a soft carpet, the lawn conceals the foundational struggle, the poisoning, the exploitation of lower social classes, and the environment beneath the promise of a vacuous and permanently stable aesthetic serenity that hides corporate greed. The greenest grass of them all: golf courses notoriously are where important business deals are made.

Strolling on a golf course is not hiking. One can sweat but the golf cart canopy, the buckets of iced water, wine, and treats are never too far away. For a certain business stratum, the golf course is a portal—an informal space where the trajectories of heads of state, CEOs, presidents, and executives can quasi-incidentally intersect into a milieu of professional informality where players chase a dimpled rubber ball. Golfing takes time—time away from screens, meetings, and distractions. Swinging only takes a few seconds. Players anxiously follow with their eyes the ball's trajectory as it cuts across the manicured landscape, through the pines, and into a sandy bunker. Then the walk and talk happen—*business.* Will players lose their temper? Will they break the rules? Talk smack behind someone's back? The game can quickly reveal important personality traits. One's responses and conduct are key to a long-lasting and fruitful business partnership. Golf does not require strenuous training and athleticism either; in that sense, it is more accessible than other sports.[8] But despite that it remains a vastly elite, male pursuit.[9] These aspects especially have more recently attracted broad criticism.

In the wake of the unprecedented droughts faced by France in 2022, Extinction Rebellion activists filled Toulouse golf course holes with concrete.[10] The following summer, as Spain sweltered under unprecedentedly scorching heat, activists dug holes in ten golf course greens and filled them with plants.[11]

The ecological impact of golf courses is real. Estimates show that the average golf course takes up roughly seventy-five hectares of land across the globe. In total, the 40,000 golf courses currently in operation collectively add up to the size of Belgium. While the summative land uptake might not seem too conspicuous, it is the resource extraction they require that raises concern. Golf courses are extremely high maintenance. Over nine billion liters of water are required in the United States each day to keep the grass green. That's nine times the amount of water used by the whole of Scotland every day. And the situation rapidly worsens in countries affected by droughts. In Thailand, a single golf course uses as much water as 60,000 rural villagers.[12] To make matters worse, 98 percent of insecticides and 95 percent of herbicides doused on golf courses reach a destination other than their target species.[13]

In a world that's rapidly and drastically being reshaped by global warming, golf turf is not only utterly unsustainable but downright unethical. As a hyper-commodified, pastoral evolution, the golf course lawn practically and metaphorically perpetuates colonial extractivist logics that destroy ecological biosystems while erasing local cultures and penalizing less-advantaged communities. From this vantage point, it seems even more paradoxical that from the Californian desert to Dubai, the most exclusive golf courses have recently sprung up in the most climatically implausible areas. While lawns of all kinds ideologically speak of a specifically human ability to subjugate nature, an even more potent brand of anthropocentric arrogance is rooted in the will to make grass grow and thrive where it technically should never exist at all.

Desert golf courses, like the one opened in 2017 by Donald Trump in Dubai, are utterly commodified, capitalist-oases—a magical and lifesaving appearance amidst utter desolation. More than any other golf course, the oasis kind brims with the miraculous strength of capitalist hubris. Its neat, green carpet, gleaming in the sweltering sun, is a live testament to capitalism's unrivalled ability to not only rule nature but also profoundly disrupt and control its biological functions and needs. These types of golf courses are a true capitalist miracle. Or, as others would have it: a deliberate insult to the idea of sustenance, a further reinforcement of elitist privilege. Corporate ideology and its view of the world as a completely disposable playground are laid bare when it reduces otherwise fertile land to a carpet upon which a small group of privileged individuals can swing a club, thereby preventing the land from feeding animals or humans or even from following its natural and ecological unravelling.

Golf course turfs are never just grass; they are living capitalist ideologies. It is therefore not a surprise that the golf turf-style lawn has also become the standard kind of greenery for corporate office parks. Situated at the edge of cities, past the outskirts, imperturbable and monumental capitalist fabulations: modular, rigorous, minimalist, neat, and banal sameness—glass boxes framed in black steel reflect the surroundings as they dematerialize into their own myth. All that matters is the brand logo floating in the upper corner. The rest is superfluous, merely structural. Universal monolithic timelessness that speaks of imperial hegemony is all the structures care to convey. Drive-by landscapes; pictures of pictures—a world of surfaces; veneers concealing abyssal, ethical hollowness. Artificially naturalized; islands of alienation surrounded by green carpets—expanded golf courses with no cup and upon which nobody even ever plays. Irremediably silent, the epitome of the superfluous. Acres and acres of sterilized land; ever so thirsty, poisoned grass that pollutes the water we drink.

But corporations haven't just fabricated their hyper-lawn aesthetic from pastoral imagery. It had already been widely adopted by other power-institutions such as universities. In the United States, Harvard (1636), William and Mary (1695), Yale (1717), and Princeton (1755) emerged from rural settings—the surrounding landscapes playing an important ideological role in augmenting the sense of elitism they embodied. These institutions thrived in the tranquil and silent countryside, away from the hustle and bustle of the city. Swaths of open land covered by closely shorn grass were intrinsically important—greens connected different buildings, allowed visibility of students and structures, and kept animals away. But in this context, the lawn also prevailed because it implicitly ventriloquized Christian values. It bolstered the spirit and restored moral order. Keen cultivation and maintenance reflected the constancy and keenness of the studious mind, its implied pragmatism and necessary clarity. Ultimately, when the first English colleges were established at Oxford and Cambridge, they adopted the monastic model rather than that of urban continental universities like Bologna or Paris.

Between 1865 and 1900, in the United States, Frederick Law Olmsted (the designer of NYC's Central Park) and his sons went on to produce thirty-six designs for college campuses all of which, in one way or another, reinforced and codified the association between the lawn and progressist enlightenment. By the 1960s, they had designed more than 200.[14] Starting there, the Judeo-Christian sets of ideologies and the power-relation matrix from which the lawn historically emerged in the United Kingdom and the United States spread across the world, providing fertile ground for corporativist thinking—a model that for nearly half a century has now come to define the way universities are managed.

Nothing to do with nature.

6 FROM THE BUFFERS, BACK TO THE GARDEN

The silence of the median lawns

"There was nothing but weeds in the middle of the road. It should look like a lawn—mowed."[1] So a Delaware farmer reacted to the state's new highway median rewilding initiative that began in 2007. Across the world, wide and endless strips of green nothingness separate highway lanes. Deadly swaths and ditches, from fields to parking lots, border off the right-hand sides of roads. Buffers, margins, and perimeters—these are the no-man's-land of capitalist pragmatism. Forever only adjacent, unloved, interstitials, liminal, they belong to everyone and no one at once. Here and there—ubiquitous, identical, utterly unremarkable green deserts. These too are hyper-lawns: they couldn't be more nondescript, literally marginal, and yet so irritatingly central to everything.

Their mere existence is the evidence of the lawn's conceptual and material uncontainability. The lawn spilled out of the suburbs, off the golf courses, through the tombstones of cemeteries, and right out of corporate parks. Under the promise of utter versatility and the predictability of aesthetic uniformity, it has claimed millions of acres of otherwise fertile land, transforming them into a carpet nobody walks, rests, or plays upon. Inaccessible: only ever visible from the windshield of speeding cars, this specific typology of lawn—perhaps more than any other—is designed to exist in an exclusively visual dimension. A picture of a picture, a blurry hue, an abstract symbol: the highway median lawn is the undeniable embodiment of our cultural addiction to soulless banality and anonymous superficiality.

Simple, straightforward, plainer than plain—the lawn has nothing to hide, it seeks no drama. It is the equivalent of an inextinguishable droning sound, the echo of an eternal hum emanating from a nuclear power station looming on the horizon. Perhaps terrifyingly so, the majority of humans find solace in this relentless desolation. This is the only reason the lawn still thrives despite the growing awareness that it is the equivalent of skin cancer on the face of the Earth.

A thorough scan of social media exchanges from Reddit to Quora, passing through lawn enthusiast forums, quickly reveals that a neat median lawn is a deadly serious matter of nationalistic and identarian pride. A subtle and yet intense field of power-laden ideologies, its neatness symbolizes the presumed authority and reliability of a state that cares for its infrastructures as well as its citizens. As soon as a skeptic dares to ask, "Why do we spend public money to mow grass between highways?" a crowd of *lawn worshippers* quickly swarms them with myriad, often absurd sounding, justifications. The sanctity of the lawn is dogmatic and so it must remain. To the adepts of the "lawn cult," anything more than short-shorn grass is a capital sin. "Broken branches and tons of leaves rolling across lanes!?" "Trees make highways dangerous!" "Grasses and shrubs collect trash and harbor animals!" "Densely planted medians impair visibility! They are a fire hazard in dry climates . . ." "Weeds are ugly." The highway median turf is the scab covering a deeply infected wound, the symptom of an incurable malaise. Its aesthetic and ideological roots feed off a subtle but all-pervasive fascist baseline that relentlessly pulses at the core of Western consciousness.

Roads are the arteries of our anthropogenic collective superorganism. The speed, the communication, the connectivity; gasoline-fueled dreams drive our desires to lead lives that can only be full in a materialistic sense. Neatness, order, pragmatism, and high functionalism squeeze through the cracks of a dark past into the form of highway median turfs. Distilled, transfigured, essentialized to the core. Living and yet dead limbs of modernist ambitions that relentlessly promised nothing else than the steady amelioration that only progress brings. A picture of positivism where the grass is always greener right here (not on the other side of the fence) where we all drive by—comfortably numb, detached from the rest of the natural world: cocooned in a glass, rubber, and steel mechanical pod. Off we drive, leaving everything behind. Linear monuments of loss and emptiness stretch to infinity across the globe in a nearly continuous and deadly embrace. A quintessential site-non-site, persistently outdated, and yet devastatingly present in its nothingness, the highway median turf is the tangible proof that the lawn has become an inescapable, human-made specific kind of, what Reza Negarestani calls, an "omnipresent planetary entity."[2]

In his non-affirmative, para-academic theory-fiction novel, the author maps and links decay, warfare, demons, and planetary demise through Middle Eastern theologies and their relation to the contemporary War on Terror. Through its speculative meanderings, *Cyclonopedia: Complicity with Anonymous Materials* sketches out the outlines of a unique perspective on materiality, suggesting that materials like oil are not just passive resources for human use but active and powerful biological, as well as political, agents. Oil thus flows at a subterranean level, often away from sight and yet all-permeating, defining political and economic global landscapes shaped by hypercapitalist drives for consumption and high mobility. As an omnipresent planetary entity, rooted in our minds as a man-made archetype, seemingly irradicable, the lawn's roots reach deep into our psyche and crawl into our fears, anxieties, and misplaced utopian hopes. Its ubiquitousness has made it invisible: How can you see something when it has become everything? At once inside and outside, a landscape feature and a state of mind.

To a degree resonating with Timothy Morton's conception of hyperobjects, the lawn exists on scales that challenge our perception and comprehension. They 'stick' to us and cannot be easily separated from the contexts in which they exist. They are simultaneously local and global and all-pervasive. This makes them difficult to pin down or fully grasp.[3]

The lawn has domesticated us. It has seduced us and jonesed us to the point of no return. How different would the world be if we had devoted the same level of intense devotion we pour on lawns to biodiverse meadows instead?

In his controversial book *The Botany of Desire*, Michael Pollan presented the compelling argument that plants have evolved to satisfy human desires in order to ensure their own survival and propagation.[4] He explored this idea through our relationships with apple trees, tulips, marijuana, and potatoes—each representing a human desire: sweetness, beauty, intoxication, and control. This theorization is a reversal of the traditional human-centric view of domestication according to which it is us, humans, who univocally play a game of subjugation. For instance, Pollan claims that corn has "domesticated" us by becoming an essential component of our food chain.

The kind of reciprocal domestication Pollan envisions seems to gain momentum, however, only truly when we interfere with biodiversity to forcefully instate monocultures. It is when forms of agri-fascism take hold that, as a representational aesthetic model, they instate themselves in our minds, that they become psychologically internalized.

The multitude, the mass, the homogeneity, the implied unity trigger a mislaid sense of reliability and trust. The intensity: a vast multitude of plants from the same species triggers a hypnotic trance. A very real sublime paradisiac endlessness, a oneiric phantom that transcends the material borders of reality. Thanks to this process of domestication, corn has managed to spread its genes far and wide across the globe by appealing to human needs and tastes. It has become a staple food in many parts of the world and a key ingredient in a wide variety of processed foods. Pollan suggests that we have been manipulated into ensuring the plant's survival and propagation across space and time. Corn has taken advantage of us, not the other way round, just as other plants entice ants and other animals to propagate and nourish them turning them into primary caretakers: a form of long-term coevolution in which symbiosis and parasitism are tightly entwined and indissoluble.

Seen from a certain, perhaps bleaker, angle, our obsession with the lawn might look more like a form of lethal parasitism similar to that which ties *Ophiocordyceps unilateralis* and carpenter ants. Known as the zombie-ant fungus, *Ophiocordyceps* takes control of the ant's mind leading it to behave in self-destructing ways that will ultimately kill the insect to benefit the fungus. Rather interestingly, research has shown that the fungus's puppeteering requires no direct connection to the ants' brains but that it might actuate via chemical secretions.[5] Lawn lovers often wax poetic on the inebriating fragrance of freshly cut grass laced with gasoline—how this fragrance fills their nostrils and minds with childhood memories of warm summers and family gatherings. So off they go— zombified—they keep pushing the mower as it slowly kills them, their children, and the planet.

A deadly silent, green refrain

The lawn is the living emblem of a wholly fictional recursive continuity that comforts and restores, lulling us into a false sense of security that capitalist forces need in order to thrive. Aesthetic repetition, relentless homogeneity, ecological flattening, its quasi-identical sameness, familiarity everywhere. The lawn is a silent, capitalist refrain that repeats itself across the planet with very few and limited variations. Multiplicities and singularities collapse into unambiguous sameness. Its self-perpetuated modularity inserts itself in the biological polyvocality of the planet as the equivalent of long pauses: story plot holes, elisions, omissions, erasures.

According to Deleuze and Guattari, the refrain marks territories, relentlessly expressing a tension that entails a process of "territorialization" and "deterritorialization." [6] The refrain helps to establish a territory of familiarity grounded in recursivity (physical, psychological, or social) by inducing a sense of safety. However, the refrain also has the potential to change the nature of the territory, to deterritorialize, and create something new and different. As a refrain, the lawn, the kind that grows in buffers especially, is a comforting reminder of the finitude that was once so dear to philosophical humanism—the illusion that a human might be an autonomous, subject that Western thinking has nurtured for the past 500 years and that is largely responsible for the ecological crisis we are currently in. The monotonous territorial motifs and landscapes that the lawn generates are silent echoes of human uncontrasted dominion: a delusion of grandeur that's already mostly vanished—spectral vestiges haunted by our inability to comprehend the utter beauty and intrinsic necessity of the collaborative more-than-human worlds we could have chosen instead of embracing colonialism when the choice presented itself toward the end of the Renaissance.

An emblem of our fragility, like patriarchy and its figments, the lawn is constantly threatened from the inside and the outside. Its seemingly imperturbable and austere minimalism is relentlessly exposed to the risk of adulteration. Any withdrawal of watering, mowing, or fertilizing will spell its demise in a matter of months or weeks depending on the time of the year and region. Centuries of relentless biodiversity suppression, the silent devastation of so much land and animal life, could be overturned by the simplest and most economical of gestures—letting go. The cultural and ideological matrix, the aesthetic allure, the silent refrain of the lawn could swiftly break into a choir of biological polyvocality once again. But could letting go ever be enough when ecological trauma is caused solely by human actions? After initial enthusiasm, the No-Mow May initiative launched by Plantlife (a British conservation charity) in 2019 started to receive criticism for its lack of long-term benefits. The idea behind No-Mow May is to allow wildflowers to bloom by simply refraining from mowing the lawn during a crucial time of the year in the life of pollinators. The campaign's simple message, "do nothing," made it easy for people to participate and contribute to a larger environmental cause.

However, the impact of No-Mow May can vary depending on the specific local environment. For example, in areas where the grass is already very short, allowing it to grow for a month may not provide much additional habitat or resource for wildlife. A month of non-mowing is just not sufficient. Also, the sudden and unmanaged interruption of mowing in most cases leads to an indiscriminate weed infestation, oftentimes only clover and dandelion will emerge—not quite the idyllic meadow of many botanical species one expected. After blooming, their seeds will spread far and wide, making it much harder to bring back the lawn to its uncontaminated uniformity in the following months. It will require more pesticides. Experts have also claimed that unshorn lawns can go into shock once they are drastically cut again as suddenly exposed to the sun, the grass' crown wilts in the summer months. That's why more productive opportunities for rewilding might lie elsewhere.[7]

More than other privately owned lawns that surround homes or institutions, buffer zones can bear the promise of new beginnings precisely because the power relations that hold them in place are weakened by their liminal essence.

Liminality, as Arnold van Gennep put it, is transitional in the sense that it always incorporates the seed of ambiguity and potential disorientation. Liminal spaces can also offer productive opportunities for openness and indeterminacy to become productive. Liminality can breed creativity.[8] Victor Turner noticed how the ambiguity of liminal spaces can also foster a sense of community and collectivity. Turner's liminal spaces are where the normal limits of thought, self-understanding, and behavior become relaxed, a place where the past is momentarily negated, suspended, or abrogated, and the future has not yet begun.[9] And according to Carl Jung, liminal spaces are psychological states of mind where the conscious and unconscious meet. In this space, individuals confront their shadow selves, leading to a process of self-discovery and transformation.[10] The thread of potential regeneration and possibility for change that runs deep into liminal areas also permeates the work of Homi Bhabha. In his view, liminality generates "third spaces"—areas for cultural congregation in which negotiation between disparate quantities can often take place—spaces of hybrid identity, generative ambiguity, and reconfigurations.[11]

With so much at stake, harnessing the generative potential of liminal lawns is not an easy task. As colonial footprints, they just can't be easily decolonized. Like the colonial erasures perpetrated by settlers can't just be simply rewritten, so the land can't be rewilded either. The silence of the lawns, in its deafening hollowness, will forever haunt the motifs and movements of the future. It won't ever be what it once was, for we don't even know exactly what it sounded like. But it will be different. As jurisprudence professor Joel Modiri argues, "Decolonisation is an insatiable reparatory demand, an insurrectionary utterance, that always exceeds the temporality and scene of its enunciation. It entails nothing less than an endless fracturing of the world colonialism created." Law professor Foluke Ifejola Adebisi further states that "we cannot decolonise while relying on colonial logics of commodification of labour and space."[12] This is also why, in all its proposed forms, the rewilding of lawns has for a few years now been a contested concept: its impossibility rooted in the "re-prefix" and its backwards-looking agency. Maybe all that is available to us are new forms of *wilding*, not rewilding?

The term "rewilding" was first used in 1990 by reporter Jennifer Foote in a story about "radical environmentalism."[13] It quickly slipped out of its original scientific domain, emerging in the humanities across multiple disciplines.[14] In essence, it implies an impossible and implicit exclusion of the human from a past that was not humanless.[15] So-called rewilding processes have often aimed to return lands to artificially "natural" states of "balanced" biodiversity. Highly artificial and carefully curated endeavors, despite the sporadic consultation with Indigenous peoples; approaches and frameworks have remained dominated by privileged white settler ideologies: ecology is relentlessly haunted by memories of the Edenic Garden.[16] Since we can't really decolonize as long as we use colonial thinking, and rewilding has no genuine past example to refer back to, it is very important to focus on creating long-lasting futures that can grow in the aftermath of global trauma

Inventive and somewhat utopian proposals that attempt to harness the liminality of highway medians have recently begun to surface. The plan to grow vegetables on medians, still in the sketching phase, is the brainchild of Product Manager Surendran Kandasamy.[17] At, at present, he has yet to elucidate how harvesting might take place at the edge of fast lanes and how constant gas emissions and polluted run-off waters might impact the quality of the food itself. Taking on a similar challenge in the United States, Troy Bishop has gone so far as to argue, while conducting the necessary calculations with care, that highway medians may be great resources for farm animal grazing.

> Sixty miles is 316,800 feet. I figured very crudely there was an easy 60 foot of grazeable median which was 316,800 feet long times 60 feet wide equalling 19,008,000 square feet divided by 43,560 feet (one acre) with a grand total being over 436 acres of New York State, eco-friendly forage, minus the speeding vehicles.[18]

It might be that these plans are simply too ambitious and that all we can hope for is to gift the land back to pollinators and as many plant species as possible, although even that has presented its challenges. Not only, as previously mentioned, locals tend to find native plants in medians and at the edges of roads unsightly, but endless stretches of native meadows might be bad news for wildlife too. In 2015, the US government passed legislation encouraging the rewilding of highway medians to the benefit of pollinators. But more than one research has revealed that a staggering number of insects find their death on our roads, either run over or crushed on windshields.[19] More recently, a two-year study showed that close to 117,000 individual insects were killed on a two-kilometer stretch of highway. Based on this data, researchers estimated that "at least 9.3 billion butterflies and 24 billion bees and wasps are killed by vehicles each year."[20] Luring them closer to speeding cars by rewilding medians might become a death sentence. Perhaps, at least for the foreseeable future, highway medians are destined to remain the province of lawns due to insurmountable practical complications. But this does not mean that we should give up on other less complex liminal lands yet.

Wilding futurities

For well over half a century, artists have been initiating *wilding futurities* across the globe—their visions and determination driven by a clear sense of unease with the ecological unbalances of their neighborhoods. These processes have perhaps not surprisingly sprung from the liminality of urban neglected buffers. Two ground-breaking artistic projects from the 1960s and 1970s surprisingly mirror the two dominant "schools of thought" on rewilding lawns that dominate today's discussions. They both emerged from a rough and tumble postwar New York and their influence quickly spread across the globe.

In 1965, Alan Sonfist, who had studied art at the School of Visual Arts in New York City, set off to recreate a scale model of a precolonial forest in a lower Manhattan abandoned lot at the northeast corner of La Guardia Place and West Houston Street. *Time Landscape* gathered vegetation that formerly grew across the area back when it was the home of the Lenape, Rockaway, and Canarsie Indians Peoples before settlers dismantled and erased their legacies.

> When it was first planted, *Time Landscape* portrayed the three stages of forest growth from grasses to saplings to grown trees. The southern part of the plot represented the youngest stage and now has birch trees and beaked hazelnut shrubs, with a layer of wildflowers beneath. The center features a small grove of beech trees (grown from saplings transplanted from Sonfist's favorite childhood park in the Bronx) and a woodland with red cedar, black cherry, and witch hazel above groundcover of mugwort, Virginia creeper, aster, pokeweed, and milkweed. The northern area is a mature woodland dominated by oaks, with scattered white ash and American elm trees. Among the numerous other species in this miniforest are oak, sassafras, sweetgum, and tulip trees, arrowwood and dogwood shrubs, bindweed and catbrier vines, and violets.[21]

For decades, Sonfist's urban mini-forest stood as a monument to the possibility of reimagining our urban spaces beyond the deadly bareness of lawns. What if by law every neighborhood of every city were required to have, tend, and maintain a plot such as this as a community endeavor? How would that change our sense of pride and ownership in urban living?

Although volunteers help to occasionally maintain *Time Landscape*, its compressed topography has over time caused tree-overcrowding that has virtually completely drowned the natural undergrowth of grasses and violets. While the artist's selection of species might reflect the original native flora, a forest is always more than the sum of its plants. Fires, animals, fungi, molds, lichens, and bacteria constantly remodel and reshape thousands of acres of land in ways that cannot be reproduced within the perimeter of a concrete jungle. Though Sonfist's work contributed to the theoretical decolonization of urban spaces, the artist's original intention has been undermined by the very nature of plant growth and its indifference to ideologies and ethics.

Almost ten years later, in 1973, artist Liz Christy updated Sonfist's vision for a new age by inventing guerrilla gardening—the now widespread practice of reclaiming an overlooked green-urban patch and cultivating it for the community. With the help of locals, she transformed a derelict lot on Manhattan's Houston St. and the Bowery in New York City. The sight of abandoned lots strewn with trash and rubble was a stark contrast to the natural beauty she believed could exist in these spaces. Christy's revolutionary ideas were strongly influenced by the principles of Land Art and Environmental Art movements of the time which sought to bring art out of the galleries and into the world, often using the landscape itself as an expressive medium. A fervent activist, the artist saw guerrilla gardening as a form of protest and resistance against the growing tyranny of increasingly more corporate impositions upon urban planning. Not only did Christy found the Green Guerrillas group, which inspired many similar activist initiatives, but she also invented the now common "seed bombs": balls made of compost, clay, and seeds, designed to be thrown into inaccessible areas to spark plant growth.[22]

Half a century later, more or less, the ideologies and concerns that brought these two artists to reenvision our relationship with green in urban settings still run deep in our discussions on the lawn's future. Sonfist's *Time Landscape* has in time unwittingly become a cautionary tale on the dangers of *performative* decolonizations and rewildings. In a city that at the time was gripped by poverty, despite its noble intentions, in hindsight *Time Landscape* appears like a misuse of precious land that could have sustained communities—a miscalculation rooted in privilege.[23] In contrast, Christy's guerrilla gardens encompassed a more holistic and pragmatic ethos. They were meant not as a monument to a bygone past (as Sonfist often referred to *Time Landscape*) but as a living and vibrant community asset. Less burdened by conservationist ideologies, grounded in the present, and more attuned to the true social and environmental needs of the neighborhood, Christy's garden at the northeast corner of the Bowery and East Houston originally featured sixty raised beds planted with vegetables. These went in first and were joined by trees and ornamental borders thereafter. But the garden didn't simply provide food. Its design was shaped by mind well-being too. A much-needed respite from the hustle and bustle of the Big Apple: pathways and benches offered opportunities to rest and contemplate.[24] Furthermore, the various flowering plant species offered sustenance for pollinators and other wild creatures that frequently struggle to flourish in heavily populated metropolitan environments like New York.

The question of aesthetics looms large upon anything we do with plants and our ingrained conceptions of wealth and decorum—the sense of what is societally deemed appropriate and dignified—play important roles in what should be allowed to grow where, when, and for what reasons. But most of us seem to have forgotten that in 1943, 20 million of so-called "Victory Gardens"—converted front and back yards as well as public greens into allotments—provided 8 million tons of food: 41 percent of all the vegetable produce at that time in the United States.[25] Cities where fertile soil is smothered by lawns could be generating sustenance instead.

Inspired by this precedent as well as by Christy's effort, in 1982, artist Agnes Denes planted two acres of wheat in downtown Manhattan, at the heart of New York's financial district, on a temporarily vacant lot worth $4.5 billion. Denes's now legendary *Wheatfields* were tended by the community and provided food for the disadvantaged. Perhaps more radically than any other urban-agricultural art project before it, Denes's showed how cities could be thoroughly reimagined as self-sustaining realities if only community building could be allowed to take precedence over corporate greed.[26] Will climate change and population growth predictions demand that our lawns must be turned into cultivated land as an urgent necessity? And if so, what could we learn from the pioneering work of these artists as well as others?

Since 1995, the artists collective Futurefarmers has revolutionized our relationship with the land and agricultural practices while reconfiguring the boundaries of artistic practice. A group of diverse, multidisciplinary artists and designers, Futurefarmers is the brainchild of Amy Franceschini, Michael Swaine, Lode Vranken, and Stijn Schiffeleers. Much of their practice is anchored in the Deleuzian notion of the *assemblage*—nonhierarchical collections of relations between heterogeneous agents that intersect, come together, and enter into relations, that organize, change, structure, and restructure with one another in ways that generate reality.[27] "Who is the other that co-determines or co-constitutes what is of importance? What provokes this pre-subjective process of communization?" the artists ask.[28] The collective has harnessed the regenerative power of art to explore and highlight the importance of farming and food production, thereby bringing these critical issues to the forefront of contemporary art discourse.

Fritz Haeg's *Edible Estates* (2005)—an activist art project—pushed the envelope further by inviting homeowners to dismantle their front lawns and replace them with mini-allotments. A public display tells the story of the garden's inception and first season of growth, and includes videos, monthly garden photo updates by a local photographer, a series of food-growing workshops, and printed brochures with local planting calendars and garden resources to inspire others to follow the example. The book *Edible Estates: Attack on the Front Lawn,* published in 2010, documents sixteen of such experimental projects established across the world. The majority of these gardens were created in collaboration with horticultural, agricultural, or community gardening organizations and commissioned by local art institutions.[29]

At around the time Haeg's book came out, Los Angeles-based fashion designer Ron Finely began to plant vegetables in dirt patches next to streets in the South Central neighborhood. Finley became concerned with the food desert in his area, where fast-food restaurants and liquor stores far outnumbered grocery stores selling fresh produce. South Central LA has some of the highest rates of obesity, diabetes, and other diet-related illnesses in the country. Finley believed that by teaching people to grow their own food, he could help them take control of their health and their lives. He saw gardening not just as a way of producing food but also as a tool for social change.

Finley's first project almost instantly triggered a citing by the City of Los Angeles for violating an ordinance that prohibited gardening in public spaces. Among other challenges was the initial lack of understanding and support from locals. But in time, "The Ron Finley Project" gathered momentum. One of the most notable accomplishments was the transformation of a barren strip of land into a thriving community garden in South Central Los Angeles. This garden not only provides fresh produce to the community but also serves as an educational hub where people learn about nutrition and urban farming. Through workshops and hands-on training, it has empowered numerous individuals, particularly youth, with the knowledge and skills to grow their own food. This has also fostered a sense of community and self-reliance. Eventually, Finley's efforts led to the amendment of Los Angeles' residential parkway landscaping guidelines, allowing residents to grow fruits and vegetables on their parkways. This legal victory was a major step in promoting urban farming and combating food deserts in the city. More recently, Finley also started a "Gangsta Garden" initiative, where volunteers, including former gang members, help transform the city's vacant land into vibrant gardens.[30] In 2019, ethnobotanist, educator, and activist T'uy't'tanat-Cease Wyss of Skwxwú7mesh, Stó:lō, Hawaiian, and Swiss descent, has set up *x̱aw's shewáy̓ New Growth* 《新生林》, a regenerative garden grounded in the principles of permaculture movements: earth care, people care, and fair share. The landscape design, which emerged on a formerly vacant lot, was arranged around a Coast Salish wave-like shape comprising Indigenous medicinal plants of the Vancouver area and the community in mind rather than classic gardening aesthetics. Geometrical motifs frequently found in Coast Salish art and design also play an important role in the reconfiguration of the narrow and elongated spaces of the garden. Wyss rightly points out that this region has been a site of cultural, social, and economic struggles for Indigenous peoples, Chinese Canadians, and African diasporic peoples. The garden is therefore thought out as an open, safe gathering space cantered around a fire pit and circular seating.[31]

Aesthetic-driven, decolonizing idealism, or pragmatic and sustainable regeneration? The current discussion around the wilding of lawns across the world seems caught between these polarities. Perhaps a sustainable way forward might be found in a calibrated entwinement of the two?

Weeds: Thriving in the shadow of power

Whether to produce food, restore native and more sustainable vegetation, or support pollinators and other earthlings, rethinking our relationship with the lawn invites us to carefully consider what it might mean to find a *compromise* in the *compromised* ecologies of the Anthropocene. While applying more pressure on our political leaders to truly and effectively engage with ecological discourses is necessary, prying open pockets of green resistance anywhere possible is within everyone's reach . No matter how small it may appear at first, every gesture of care toward our earthly kin could go a long way in the not-too-distant future.

As artists show us, while urgent and effective actions are needed in order to mitigate the negative impacts of lawns across the globe, a longer-term process of aesthetic and ideological re-education is also essential to weaning us off our lawn addiction. Centuries of representational tyranny, grounded in colonialist ideologies, have predisposed us to succumb to the lure of the lawn. The aesthetic appeal of these plain and homogenous swaths of land that have come to symbolize status, power, peacefulness, health, and safety is part of a shared cultural unconscious that will require time to rewire. Art, design, and architecture will all play a crucial role in this process. In pre-photographic times, when people traveled much less than we do now, it was the canvases of artists like John Constable, Antoine Watteau, Canaletto, and John Varley, among others, that cemented the lawn's status as the ultimate status symbol of the superrich's power. As Tom Stoppard wrote in *Arcadia*: "English landscape was invented by gardeners imitating foreign painters who were evoking classical authors. The whole thing was brought home in the luggage from the grand tour."[32] So today contemporary art and popular culture imagery are crucial in reconfiguring our aesthetic preferences. The plant blindness that affects most of us can be cured, one glance at a time.

Recently, artists—who by essence are experts in aesthetics— have invited us to look more closely and carefully at buffer spaces that we thought bore no potential. In 1997, vegetal art pioneer Lois Weinberger enabled the wilding of disused train tracks in a train station—ripe with vegetal potential, a liminal space, caught between the architectural strictures of a utilitarian infrastructure and the economic forces that declare obsoleteness. Plants that grow across train tracks exist first and foremost as a statement of ecological/vegetal rights. In *Das über die Pflanzen ist eins mit Ihnen*, plants grow upon the ground they should have rightly inhabited all along— their reclaiming is political. Positioned in a high-pedestrian traffic area, the uncanny aura of Weinberger's piece ignites curiosity.[33] Why are these plants allowed to take over this specific set of tracks? Which pollinators or other animals now thrive there? How are these plants going to change throughout the seasons? If so much diversity can flourish among train tracks what could this look like in my backyard?

Wildflowers of all kinds tend to be smaller and less showy than cultivars. In a train station where people everyone is in a rush to catch a train or get to work on time, the unusual sight of wildflowers might stop people in their tracks and invite them to look more closely at their often-understated beauty. This is where our re-education of the gaze can begin. Small flowers ask us to get close and establish a more intimate connection. They awaken our senses by inviting us to smell. We need to see alternatives in order to choose alternatives. If all we see is lawns, lawns is all we will desire.

Unconcerned with fictitious notions of purity, Weinberger's installation also deliberately mixed native vegetation with neophytes. A train station is a migratory site of departures and arrivals. The species mix gestures to the all-important idea that our world is never still and that its vibrant diversity is the result of many journeys and unplanned encounters. National identity, history, ecology, and diasporic experiences: this diversity is a form of beauty rooted in the spontaneity of unrepeatable moments, all different, all essential.

The act of retraining our gaze to appreciate a broader range of vegetal forms and colors has been an important principle in the work of Zachari Logan whose exquisitely detailed, large-scale pastel drawings often portray luxuriant tangles of weeds. The artist often focuses on liminal spaces like ditches—neglected and marginal swaths of land that have historically been reclaimed by the LGBTQAI+ community. Like weeds, undesired, excluded from the normative rationality of formal gardens and manicured lawns, queer groups have resisted and regenerated in liminal places over time, staking important cultural territory and visibility. There is sheer beauty of an unconventional kind in Logan's vision. The artist's imagery subtly overlays diversity and biodiversity as interrelated forms of inclusive and essential generativeness that the eco-fascism of the lawn has for centuries smothered or even successfully eradicated.

The association between the lawn and systems of normative social repression has more recently drawn many artists to creatively engage with weeds: the lawn's sworn enemy. Weeds are now hyper-political, encompassing important and nuanced discourses that link biology, ecology, and geography. To contemporary artists, a weed no longer is just "a plant out of place," as Jim Thompson declared in his 1952 novel *The Killer Inside Me*, but a resilient and resistant, active player, capable of unsettling hierarchies and thus endowed with the potential of changing our world. Contemporary artists show us that weeds actively weaken the foundations of order and perceived conceptions of purity; they thrive in the shadow of power. They perturb cultural and social normativity by relentlessly refusing to stay put and comply. They invite us to craft new registers of resistance modeled on ecological practices of tenacity and dissidence grounded in a past of injustice that still haunts the present.

Eventually, artists working with weeds hope to make the term redundant. A sign that things might be heading in that direction came in April 2023, as the iconic London Chelsea Flower Show officially rebranded weeds "hero plants," dedicating their prestigious displays to brambles, thistles, and knapweed among others. Landscape and garden designer Tom Massey said:

People often get really stressed out about dandelions on the lawn but they are a good source of nectar [for bees] because they flower early in the year and they're really nice to look at. Even foxgloves and knapweed [are welcome]. There's a lot of stigma around the word "weed." [34]

Institutional statements such as these are important for they might, in time, rewire our outmoded conceptions on a popular scale. However, as we wait for that change to happen, nothing stops us from reinventing the way we garden—now. How much lawn do we really need in our yards? Can we compromise to ensure that wildlife can also thrive in the open spaces we share? Can we reduce the amount of water and pesticides we pour on our lawns? These questions might sound new to some, but invitations to carefully consider how much lawn should one have in their yard were issued already in the nineteenth century, as the lawn gained popularity. In his 1870 book *The Wild Garden*, William Robinson tried to instil a slow but steady revolution: "Surely it is enough to have a portion of lawn as smooth as a carpet at all times, without sending the mower to shave the 'long and pleasant grass' of the other parts of the grounds."[35] It is not too late to embrace his invitation. Any move toward caring for the land, plants, and other earthlings—no matter how small— will make a difference.

The end of the lawn: From the ground up

In truth, it is not easy to wild a lawn. It is certainly not as easy as doing nothing. In a controversial article titled "Ditch Your Spades, Forget Fertilizers, Listen to the Weeds: Alys Fowler's Guide to Laid-back Gardening" published in 2023, the author invites readers to let things go.[36] Past the somewhat alarming headline that triggered fierce criticism from some of the most well-respected British garden gurus, Fowler offered some good advice. She proposed a more slow-phased and intimate relationship with the soil; invited readers to read the predominance of some plants and the absence of others as telling signs of mineral unbalances; and she also asked us to slow down and interact with land parsimoniously rather than aggressively—to follow rather than lead. Giving up nitrogen and phosphorous-based fertilizers—one of her key recommendations—can also play an important role in reeducating our attention, observation, and patience. Gardening should be an opportunity for us to slow down and focus while the world constantly asks us to be in multiple places at once. Fertilizer brands like Miracle Grow have miseducated us to want bigger plants as fast as it is "naturally" possible. These fertilizers often end up depleting the soil of stored carbon and saturating it with salts that quickly kill fungi beneficial to plant life. Ultimately, Fowler invited us to embrace promiscuity by letting plants genetically mix rather than obsessing about the fixity of color pallets and the neat demeanor of cultivars in order to strengthen their genetic pools.

In all of this, I am not advocating giving up on gardening, but shifting the perspective on what needs doing. If the dandelion, dock or bramble isn't in the way, leave it. If the plant goes down in an orgy of aphids, leave it for some other garden being to clear up. Let plants die in place, learn to watch and observe before you make a move.

You'll see that nature is way more willing to help than cause trouble.[37]

While Fowley's laid-back gardening model might require a bit more work, or at least close monitoring, than the author might be willing to admit, it certainly offers a valuable opportunity to rethink our gardening practices and aesthetics from the ground up. Wilding is not about removing human interaction with the soil in order to return to a natural state of utopian purity—our gardens can never truly replicate the complex ecological tapestry of wild prairies and forests, some of which have not been substantially disturbed for 8,000 or 12,000 years. The ecological stability of these places is the result of intimate biological, geological, and evolutional entwinements. Replacing cultivars with nativars—the pseudo-wild selected varieties on sale at box-chain stores—will not lead to a longlisting difference. Radical re-education is what's needed. In order to spell the end of the lawn and substitute it with sustainable alternatives, we must first give up any notions of purity.

For many years now, artist Egle Oddo has created *evolutionary gardens* as living public sculptures. Oddo indiscriminately plants native and cultivar species together to initiate a long-term process of coevolution. Wholly deregulated, her gardens confound the distinction between site for agriculture and site for leisure, and wilderness.[38] As a result, different species or genera might eventually naturally hybridize or generate unusual biotic assemblages—in this case too, the artist is renegotiating her agency in a process of co-creation with plants and land. Oddo is careful to position herself in this relationship not as a "steward"—a term often borrowed from Indigenous identities that in the hands of others can quickly become problematic—but as a collaborator. Her approach, in her words, generates "performative habitats." "I intercept and join in the larger action performed chorally by life forms present in a given habitat."[39] Engaged approaches designed to sustain ecosystems also invite us to negotiate mindfulness and empathy as much as engage with the complexities of historical pasts and their legacies.

In 2022, over the span of eighteen months, poet and artist Precious Okoyomon installed and tended to a highly diverse rooftop garden at the Aspen Art Museum in Colorado. Freely including plants traditionally considered invasive as well as natives, the artist invited viewers to consider the inescapable paradoxes of a world that has been radically reconfigured by the atrocities and exploitations of colonialist pasts. Okoyomon's garden called *Every Earthly Morning the Sky's Light Touches Ur Life Is Unprecedented in Its Beauty* is a unique blend of poetry, performance, and installation. The space isn't simply a static display for aesthetic contemplation but a living, ever-evolving, multispecies organism that reflects the artist's belief in the transformative power of nature. Okoyomon has inscribed poems on rocks, embedding them in the soil or reciting them during live performances. Their garden has shifted the emphasis on process and change as defining forces we must embrace in order to better grasp and negotiate the complexities of past histories that still shape our present, a reflection of the fragility and beauty of life in all its unending flux. As a monoculture of non-native grass, the lawn hinders our commitment and defers our responsibility to amend a fraught past of exploitation of both, the ecological histories of the landscape and the lives of peoples, plants, and animals that have for centuries forcefully become enmeshed in it.

A warning: despite its regenerative promises, *wilded gardens* will most often look messy. They know no true aesthetic order. That's where their beauty lies. But they also won't look good in a traditional sense, all-year round either (not even all summer!). A beautiful meadow might peak in July, from then on it will be yellow grasses and dray stalks all the way to the first frost. Throughout, blooming might be intermittent with long swaths of browning leaves in between. Over centuries, cultivars like petunias, geraniums, impatiens, and other patio favorites have been engineered to generously bloom all season. These impossible hyper-plant aesthetics have largely skewed and distorted our aesthetic taste and expectations. Centuries of intensive gardening approaches have given birth to calendar models designed to synchronize plant life and give us what we want all spring and summer long: boundless amounts of flowers and lush foliage, always, throughout—no compromise.

Up until recently, that's how a professional gardener would have thought about the planting of different varieties in the same space. A pyrotechnic display of continuous alternations of color bursts followed by a grand autumnal finale. These gardens have been systematically dreamed up and planted for us without pollinators in mind. In fact, oftentimes, the well-calibrated unravelling of carefully planned aesthetic-gardens often requires their exclusion. Today, untangling our aesthetic preferences from the rhetorical structures of classical art and architecture, contemporary artists promote new models of inclusivity, empathy, and acceptance that bear the potential of revolutionizing our relationship with gardening and by extension with the land. The messiness, the downtime, the imperfections—wilded gardens know a different kind of all-year-round beauty that reaches deep into the soil and spreads far into the journeys of the creatures that will visit it.

Compromise, collaboration, and reciprocity

Right in the heart of Chicago, above an underground parking lot, thrives Lurie Garden: a two-and-a-half acre stretch of land that, since 2004, has been the home of many native and non-native plants. Described as "an urban model of responsible horticulture, providing a healthy habitat for a wide variety of plants and wildlife," over the span of a few years, the garden has developed into an urban sanctuary for insect and bird life.[40] Surrounded by some of the tallest skyscrapers in the world, Lurie Gardens is a true biodiversity gem designed to reeducate our aesthetic taste and reconfigure our ecological conceptions. Here, there's no room for lawns. The nearby lawn, opposite the dramatically sublime Pritzker Pavilion designed by Frank Gehry, is off-limits during the day. Lush and trim, it is reserved for audiences attending summer evening performances. Lurie Garden is for all earthlings, the more-than-human world first.

Whether domestic, institutionally owned, or public, urban greens are some of the hardest to wild because of legislations and the persistent notion that untidy vegetations in cities equate to a negligent government. While all flowerbeds embellishing the nearby Magnificent Mile (the city's premier commercial district) are regularly turned over every season—tulips and daffodils in spring, tropicals in summer, and chrysanthemums in the autumn—Lurie Gardens is allowed to flow with time. In late October, as the first frost arrives, its keepers don't cut anything down. While armies of garden maintenance companies descend upon private gardens, chopping, mulching, and blowing everything that was once green, Lurie Gardens basks in the austere solemnity of bronze stems and golden leaves—a vegetal architectural triumph: the skeletal memory of summer exuberance. Unscathed by the insult of Halloween and Christmas decorations, so it remains all winter: textures, hues, lines, shapes, subtle contrasts, infinitely rich earthly palettes. Lurie Gardens speaks a vegetal language that most of us city dwellers have been taught to ignore or even despise.

But for years, at the beginning of spring, Lurie Gardens was also mechanically mowed to the ground so that crocuses and other early spring flowers could bloom unencumbered.

As the Marketing-Communication Officer said, "An array of tulips wouldn't be as Instagrammable mixed in with a bunch of twigs sticking up from the ground."[41] However, in 2018, the garden's director, Laura Ekasetya, came to the conclusion that the spring mowing did not fully align with the garden's ecological mission. Throughout the year, a number of insect species, carpenter and mason bees as well as other insects overwinter, either as adults or as chrysalids, build nests in the lower sections of hollowed-out plant stems. To save the majority of insects that live in the garden, strategically selected areas were designated to receive no mowing or to only be hand pruned to leave the lower part of plants untouched. No-mow areas have been expanded every year since. Engaging signage educates visitors about this important measure.

Lurie Gardens is in essence an iteration of what Anna Tsing calls a "site of slow disturbance" which "refers to anthropogenic ecosystems in which many other species can live. Slow disturbance landscapes are those that nurture interspecies collaborations. They are not untouched by the presence of humans, the ultimate weedy invader."[42] These landscapes, which are in a state of constant becoming, foster a kind of contaminated diversity capable of generating cosmopolitan kinship.

Compromise, collaboration, and reciprocity. Potawatomi professor of environmental biology Robin Wall Kimmerer invites us first and foremost to "build good soils."[43] "Soil, to me, is a worthy ancestor for it is simultaneously the repository of what has come before and the garden for what is yet to come. It joins the realm of memory to the ultimate source of becoming."[44] According to Kimmerer, building good soil is a responsibility we all have toward the planet and future generations. Good soil, she says, holds ideas. "Biodiversity is the imagination of the earth; it propels the emergence, the evolution of our ascendants. Building good soil means preserving a world rich enough in biodiversity that it can imagine itself anew."[45] This is what the lawn denies us all, other earthlings included. In its monocultural dumbness, it prevents the world from imagining itself anew. Stuck in its own silent past, perpetuated by herbicide and mowers, it can no longer grow into new futurities.

Instead of silencing the soil, we need to relearn how to listen to it, how to engage in conversations. We must begin to think with the soil anew. We must acknowledge, first and foremost, that in more than one way, the land is not ours because it came before us and because it will outlive us. Rethinking the very conception of land ownership will prove essential. Which land is owned in ways that always simultaneously entail the presence and benefit of others? Which land should never be wholly considered "private" even if it is sold under such terms in a legal agreement?

Gardens and green areas of all types should be regarded as perpetually belonging to all earthlings. The land that surrounds our architectures should never fully become one's private property in the sense that it always is, more than any other indoor space, a communal life-crux, a shared, ancestral crossway toward which we have first and foremost responsibilities: to keep it fertile, healthy, rich, and diverse for all—to retain its essential livingness intact, to foster it, and always, at all costs, to avoid spoiling it. Over the last century, we have lost sight of this very important contract with the more-than-human world and have rolled lifeless carpets of green out of our homes and onto the land that also belongs to the plants, the bees and butterflies, the birds and the ants, and all the other critters. The lawn is the quintessential manifestation of our anthropocentric selfishness—its monotonous, exclusionary flatness is a testament to the human egoism that is relentlessly impoverishing our world. Let's reeducate ourselves. Let's rethink the boundaries and borders, the ins and outs, the now and forever. Let's free the land. Let's give every green space the chance to engage in continuous and unpredictable polyamorous encounters and exchanges, shared growth, endless enrichment, and earthly coevolution grounded in compromise, collaboration, and reciprocity. Let's build good soil, together, so that the earth can begin to think again.

That's all about nature, first and foremost.

Us included.

BIBLIOGRAPHY

Adler, Ben. "The Case against Golf." *The Guardian,* June 14, 2007.

Alighieri, Dante, and Musa, M. *Dante's Inferno: The Indiana Critical Edition.* Indiana University Press, 1995.

Allen, Will, and Haeg, Fritz. *Edible Estates: Attack on the Front Lawn: A Project by Fritz Haeg.* Metropolis Books, 2010.

Aloi, Giovanni, ed. "Beyond Morphology," special issue. *Antennae: The Journal of Nature in Visual Culture,* no. 18 (2011).

Aloi, Giovanni, ed. "Why Look at Plants?" *Antennae: The Journal of Nature in Visual Culture,* no. 17 (2011).

Aloi, Giovanni, ed., and principal au., *Why Look at Plants? The Botanical Emergence in Contemporary Art.* Brill, 2018.

Aloi, Giovanni, and Susan McHugh. *Posthumanism in Art and Science.* Columbia University Press, 2021.

"Anarchist Golfing Association Trashes Grass-Seed Research Facility." *The Record-Currier,* December 20, 2001. https:// www.recordcourier.com/news/2001/dec/20/anarchist-golfing -association-trashes-grass-seed-r/.

Anderson, Dave. "It's Time to Clean Up the War Vocabulary in Sports." *New York Times,* March 23, 2003. https://www.nytimes .com/2003/03/23/sports/sports-of-the-times-it-s-time-to-clean -up-the-war-vocabulary-in-sports.html.

Anderson, Kat. *Tending the Wild: Native American Knowledge and the Management of California's Natural Resources.* University of California Press, 2013.

Arjmand, Reza. *Public Urban Space, Gender and Segregation.* Routledge, 2016.

Barad, Karen. *Meeting the Universe Halfway: Physics and the Entanglement of Matter and Meaning.* Duke University Press, 2007.

Benedictow, Ole Jørgen. *The Complete History of the Black Death.* Boydell & Brewer, 2021.

Bengtsson, J., J. M. Bullock, B. Egoh, C. Everson, T. Everson, T. O'Connor, P. J. O'Farrell, H. G. Smith, and R. Lindborg. "Grasslands—more important for ecosystem services than you might think." *Ecosphere* 10, no. 2 (February 2019). https://doi.org/10.1002/ecs2.2582.

Benjamin Heim Shepard. *Sustainable Urbanism and Direct Action.* Rowman & Littlefield, 2014.

Bernstein, Mark. F. *Football: The Ivy League Origins of an American Obsession.* University of Pennsylvania Press, 2001.

Bhabha, Homi K. *The Location of Culture.* Routledge, 2012.

Billson, Chantelle. "Study reveals sports most notorious for homophobic and transphobic abuse." *PinkNews*, April 22, 2023. https://www.thepinknews.com/2023/04/22/sports-football-cricket-homophobic-transphobic-abuse/.

Bishop, Troy. "Median Grazing: Crazy Enough to Work?" *Country Folks*, December 20, 2013. https://countryfolks.com/median-grazing-crazy-enough-to-work/.

Bonnett, Alistair. *Off the Map.* Islington. Arum Press Ltd, 2014.

Bormann, F. Herbert, Diane Balmori, and Gordon T. Geballe. *Redesigning the American Lawn: A Search for Environmental Harmony.* Yale University Press, 2001.

Braidotti, Rosi. *The Posthuman.* Cambridge: Polity, 2013.

Bryant, Levi, Nick Srnicek, and Graham Harman. *The Speculative Turn.* RePress, 2011.

Burke, Edmund. *A Philosophical Inquiry into the Origin of Our Ideas.* Simon and Schuster, 2012.

Butler, Judith. *Gender Trouble: Feminism and the Subversion of Identity*. Routledge, 1999.

Cane, Jonathan. *Civilising Grass*. Wits University Press, 2019.

Carrington, Damian. "Car 'Splatometer' Tests Reveal Huge Decline in Number of Insects." *The Guardian*, February 12, 2020. https://www.theguardian.com/environment/2020/feb/12/car-splatometer-tests-reveal-huge-decline-number-insects/.

Carroll, Valerie Padilla. *Who Gets to Go Back-To-The-Land?* University of Nebraska Press, 2020.

Chan, Wilfred. "Central Park Warned a Festival Would Ruin Its Lawn. New York Went Ahead with It." *The Guardian*, October 6, 2023. https://www.theguardian.com/us-news/2023/oct/06/central-park-great-lawn-global-citizen-festival.

Clarke, P.A. 2014 "The Aboriginal Australian cosmic landscape, Part 1: The ethnobotany of the Skyworld." *Journal of Astronomical History and Heritage*, 17, no. 3 (2014): 307–35. http://dx.doi.org/10.3724/SP.J.1440-2807.2014.03.05.

Cluff, Jeffrey. "Arizona Cardinals' field slammed again as 'atrocious,' 'deplorable,' 'embarrassing.'" *The Arizona Republic*, August 14, 2023. https://www.azcentral.com/story/sports/nfl/cardinals/2023/08/14/arizona-cardinals-field-turf-nfl-playing-surface-slammed-again/70589468007/.

Cluff, Jeffrey, and Lorenzo Reyes. "'The Sodfather' George Toma explains why Super Bowl 57 field was so slippery." *USA Today*, February 28, 2023. https://www.usatoday.com/story/sports/nfl/super-bowl/2023/02/28/super-bowl-57-slippery-field-conditions-explained/11368392002/.

Coakley, Jay. *Handbook of Sports Studies*. Sage, 2010.

Council on Foreign Relations. "Global Conflict Tracker." Accessed January 4, 2024. https://www.cfr.org/global-conflict-tracker.

Cronon, William. "The Trouble with Wilderness: Or, Getting Back to the Wrong Nature." *Environmental History,* 1.1 (1996), 7–28.

Curry, Graham., and Eric Dunning. *Association Football.* Routledge, 2015.

Davies, Chris. "The High-Tech 2-Year Story Behind Super Bowl 2023's Disastrous Turf." *SlashGear.* February 13, 2023. https://www.slashgear.com/1197410/the-high-tech-2-year-story-behind-super-bowl-2023s-disastrous-turf/.

Deleuze, Gilles and Felix Guattari *A Thousand Plateaus: Capitalism and Schizophrenia.* Translated by B. Massumi. Bloomsbury, 2017.

Dickson, Kiana. "What is the anti-lawn movement? Expert opinions and how you can join in with the change." Better Homes & Gardens, September 7, 2022. https://www.homesandgardens.com/gardens/what-is-the-anti-lawn-movement.

Diep, Francie. "Lawns vs. crops in the continental U.S." *Scienceline,* September 19, 2011. https://scienceline.org/2011/07/lawns-vs-crops-in-the-continental-u-s/.

Dosh, Kristi. "Golfers Make Better Business Executives." *Forbes,* May 16, 2016.

Duggal, Anna, and Ali Marium. "Why do more than 800 million people live in hunger?" Al Jazeera, May 23, 2023. https://www.aljazeera.com/news/2023/5/28/why-is-global-hunger-on-the-rise-2.

Fellows, James. "Leaf blower Legislation: The *WaPo* and the *Current* Weigh In." *The Atlantic,* January 18, 2016. https://www.theatlantic.com/national/archive/2016/01/leafblower-legislation-local-press/625260/.

Fest, Racheal. "Culture and Neoliberalism: Raymond Williams, Friedrich Hayek, and the New Legacy of the Cultural Turn." *Mediations: Journal of the Marxist Literary Group* 34, no. 2 (Spring 2021): 9–27.

Feuer, Alan. "Settlement on Use of Central Park's Great Lawn." *New York Times*, January 8, 2008. https://www.nytimes.com/2008/01/08/nyregion/09cnd-lawn.html.

Firestone, Jared. "Artificial Turf Cancer Lawsuit: Toxic Compounds May be Linked to Cancer." Expert Institute, May 21, 2024. https://www.expertinstitute.com/resources/insights/artificial-turf-cancer-lawsuit-toxic-compounds-may-be-linked-to-cancer/.

Folukei, Fejola. "Why I Say 'Decolonisation Is Impossible.'" *Foluke's African Skies*, December 17, 2019.

Foote, J. Jennifer "Radical environmentalists are honing their militant tactics and gaining followers." *Newsweek* 115 (1990): 24.

Fort, Tom, and Georges Teyssot. *The American Lawn*. Princeton Architectural Press with Canadian Centre for Architecture, 1999.

Fort, Tom. *The Grass is Greener*. Harper Collins, 2008.

Foucault, Michel. *Discipline and Punish: The Birth of the Prison*. 1975. Reprint, New York: Vintage Books, 1995. Translated by Alan Sheridan.

Fowler, Alys. "Ditch Your Spade, Forget Fertiliser, Listen to the Weeds: Alys Fowler's Guide to Laid-Back Gardening." *The Guardian*, May 10, 2023. https://www.theguardian.com/lifeandstyle/2023/may/10/ditch-your-spade-forget-fertiliser-listen-to-the-weeds-alys-fowlers-guide-to-laid-back-gardening.

Futurefarmers. "Futurefarmers." Accessed January 4, 2024. https://www.futurefarmers.com/.

Gagliano, Monica. *Thus Spoke the Plant: A Remarkable Journey of Groundbreaking Scientific Discoveries and Personal Encounters with Plants*. Penguin, 2018.

Gardner, David. "No Mow May in Full Swing... so What's Wrong with 'No Mow May'?" Buckeye Yard & Garden Online, May 12, 2023. https://bygl.osu.edu/node/2137.

Ghosh, Amitav. *The Great Derangement: Climate Change and the Unthinkable*. The University of Chicago Press, 2016.

Gibson, Terrill L. *The Liminal and the Luminescent*. Wipf and Stock Publishers, 2021.

Gowdy-Wygant, Cecilia. *Cultivating Victory: The Women's Land Army and the Victory Garden Movement*. University of Pittsburgh Press, 2013.

Graafland, Arie. "The Dance of Versailles: Nature, Circe, and the Garden." In *Earth Perfect: Nature, Utopia, and the Garden*. Black Dog Publishing, 2012.

Gregory, Andy. "Leaf blowers contributing to 'insect armageddon' and should be avoided, German government warns." *The Independent*, November 19, 2019. https://www.independent.co.uk/climate-change/news/leaf-blowers-insect-noise-pollution-germany-gardening-a9208366.html.

Grierson, Jamie. "Chelsea Flower Show Gardens to Rebrand Weeds as 'Hero' Plants." *The Guardian*, April 25, 2023. https://www.theguardian.com/lifeandstyle/2023/apr/25/chelsea-flower-show-gardens-rebrand-weeds-hero-plants.

Hall, Matthew. *Plants as Persons: A Philosophical Botany*. SUNY Press, 2011.

Hamacher, Duane, and Kirsten Banks. "The Planets in Indigenous Australian Traditions." In *Oxford Research Encyclopedia of Planetary Science*, edited by Peter Read. Oxford University Press, 2018.

Haraway, Donna. *Simians, Cyborgs, and Women: The Reinvention of Nature*. Routledge, 1990.

Harman, Graham *The Quadruple Object*. Zero Press, 2011.

Harman, Graham. "Art and Objecthood." In *Realism Materialism Art*. Edited by C. Cox, J. Jaskey, and S. Malik. Sternberg, 2015.

Harman, Graham. *Third Table*. In *100 Notes, 100 Thoughts: dOCUMENTA*. Hatje Cantz, 2012.

Harris, Stephen. *Grasses*. Reaktion Books, 2014.

Haslam, S Alexander, Katrien Fransen, and Filip Boen. *The New Psychology of Sport and Exercise*. SAGE, 2020.

Hayles, N. Katherine. "Unfinished Work: From Cyborg to Cognisphere." *Theory, Culture, and Society* 23, no. 7–8 (2006): 159–66.

Heckscher, Morrison H. *Creating Central Park.* Metropolitan Museum of Art, 2008.

Herring, Mary. "From Sheep to Robots: The History of Lawn Mowers–Iron Solutions." *Ironsolutions.com*, 2020.

Heywood, Paolo. "Anthropology and What There Is: Reflections on 'Ontology.'" *Cambridge Journal of Anthropology* 30, no. 1 (2012): 143–51.

Heywood, Paolo. "The Ontological Turn." In *Cambridge Encyclopaedia of Anthropology*, edited by Felix Stein. Cambridge University Press, 2017.

Huxley, Anthony. *An Illustrated History of Gardening.* Paddington Press LTD, 1978.

IPCC. "Summary for Policymakers of IPCC Special Report on Global Warming of 1.5°C approved by governments." October 8, 2018. https://www.ipcc.ch/2018/10/08/summary-for -policymakers-of-ipcc-special-report-on-global-warming-of-1 -5c-approved-by-governments/.

Jenkins, Virginia Scott. *The Lawn: A History of an American Obsession.* Washington DC: Smithsonian Institution Press, 1999.

Johnson, Carla K. "What does the science say about the grass vs. turf debate in sports?" *AP News*, October 9, 2023. https:// apnews.com/article/nfl-aaron-rodgers-achilles-grass-artificial -turf-79212f5443cd2a0d30fe8c9d981b13c0.

Jones, Sam. "Extinction Rebellion Plugs Holes on 10 Spanish Golf Courses in Water Protest." *The Guardian,* July 3, 2023. https:// www.theguardian.com/world/2023/jul/03/extinction-rebellion -plugs-holes-on-10-spanish-golf-courses-in-water-protest.

Jørgensen, Dolly. "Rethinking Rewilding." *Geoforum* 65 (2015): 482–88.

Junior, Emanuel Leite and Rodrigues, C. *China, Football, and Development*. New York: Taylor & Francis, 2023.

Kearns, C. A., and L. Prior. "Toxic greens: a preliminary study on pesticide usage on golf courses in Northern Ireland and potential risks to golfers and the environment." *WIT Transactions on The Built Environment* 134 (2013): 173–82.

Kendler, Jenny. "An Open Book of Grass." In *Why Look at Plants? The Botanical Emergence in Contemporary Art*, edited by Giovanni Aloi. Brill, 2019.

Kenna, Mike. "A Turfgrass Timeline: The History of Creeping Bentgrass Breeding." *USGA Green Section Record* 62, no. 1 (2024). https://www.usga.org/content/usga/home-page/course-care/green-section-record/62/issue-01/a-turfgrass-timeline--the-history-of-creeping-bentgrass-breeding.html.

Killcoyne, Hope Lourie, and Mary Lee Majno. *The Lost Village of Central Park*. Silver Moon Press, 1999.

Kimmerer, Robin Wall. *Braiding Sweetgrass: Indigenous Wisdom, Scientific Knowledge, and the Teachings of Plants*. Milkweed Press, 2013.

Kingsley, Thomas. "Climate Activists Fill Golf Course Holes with Cement." *The Independent,* August 14, 2022. https://www.independent.co.uk/climate-change/news/climate-activist-france-water-ban-b2144729.html.

Kohn, Eduardo. *How Forests Think: Toward an Anthropology Beyond the Human*. University of California Press, 2015.

Kourik, Robert, and R. Creasy. *Roots Demystified: Designing and Maintaining Your Edible Landscape Naturally*. Metamorphic Press, 1986.

Kubiak, Jo ana. "Cutting Back on the Cut-Back." *Lurie Garden,* March 15, 2019. https://www.luriegarden.org/cutting-back-on-the-cut-back/.

LaPier, Rosalyn. "Indigenous Knowledge, Grasslands and Bison." *The American Buffalo: A Film by Ken Burns*, October 16,

2023. https://www.pbs.org/kenburns/the-american-buffalo/
indigenous-knowledge-grasslands-and-bison.

Lauletta, Tyler. "Eagles and Chiefs players both agreed that the
Super Bowl's $800,000 turf was terrible: 'It was like playing on
a water park.'" *Business Insider*, February 13, 2023. https://www
.businessinsider.com/eagles-chiefs-super-bowl-field-turf-bad
-slip-water-park-2023-2.

Levitan, Mark, and Susan Wieler. "Poverty in New York City,
1969–99: The Influence of Demographic Change, Income
Growth, and Income Inequality." *SSRN Electronic Journal* 14,
no. 1 (2008). http://dx.doi.org/10.2139/ssrn.1185602

Light, Jonathan Fraser. *The Cultural Encyclopaedia of Baseball*.
McFarland and Company, 2005.

Lurie Gardens Stewardship. "Sustainability." Accessed October 10,
2024. https://www.luriegarden.org/about/sustainability/.

Lynch, David, director. *Blue Velvet*. Warner Brothers, 1986. 120
min.

Lynch, Patricia Ann, and Jeremy Roberts. *African Mythology A to
Z*. Chelsea House, 2010.

MacKay, Robin, Luke Pendrell, and James Trafford. *Speculative
Aesthetics*. Urbanomic, 2014.

Mancuso, Stefano, and Alessandra Viola. *Brilliant Green: The
Surprising History and Science of Plant Intelligence*. Island Press,
2015.

Marder, Michael. *Plant-Thinking: A Philosophy of Vegetal Life*.
Columbia University Press, 2013.

Merleau-Ponty, Maurice. *The Structure of Behavior*. Translated by
Aldin Fischer. Metheun, 1965.

Meredith, Martin. *Mandela*. Simon and Schuster, 2010.

Milman, Olivia. "Tree-mendous news: noisy gas-powered leaf
blowers banned in Washington DC." *The Guardian*, January 5,
2022. https://www.theguardian.com/us-news/2022/jan/05/gas
-leaf-blowers-banned-washington-dc.

Miran, Lana. "How Artist T'uy't'tanat-Cease Wyss Uses Ethnobotany to Reconnect Unceded Land with Indigenous Culture." *Arts Help,* September 20, 2023. https://www.artshelp.com/how-artist-tuyttanat-cease-wyss-uses-ethnobotany-to-reconnect-unceded-land-with-indigenous-culture/.

Morgan, Tyne. "Artificial Turf Made with Soybeans Is Growing in Popularity, Now on Display at the San Diego Zoo." *The Scoop,* March 14, 2023. https://www.thedailyscoop.com/news/retail-industry/artificial-turf-made-soybeans-growing-popularity-now-display-san-diego-zoo.

Mortimer-Sandilands, Catriona, and Bruce Erickson, eds. *Queer Ecologies: Sex, Nature, Politics, Desire.* Bloomington: Indiana University Press, 2010.

Morton, Timothy. *Ecology without Nature: Rethinking Environmental Aesthetics.* Harvard University Press, 2007.

Morton, Timothy. *Hyperobjects: Philosophy and Ecology after the End of the World.* University of Minnesota Press, 2017.

Mozingo, Louise A. *Pastoral Capitalism.* MIT Press, 2016.

Murray, A.T. *Homer, The Odyssey with an English Translation.* Harvard University Press, 1991.

Nagel, Thomas. "What Is It like to Be a Bat?" *The Philosophical Review* 83, no. 4 (1974): 435–50. https://www.jstor.org/stable/2183914.

National Golf Foundation. "Golf Research and Industry Data." Golf Industry Facts. Accessed January 4, 2024. https://www.ngf.org/golf-industry-research/.

Vivekanandhan, Neelambari, and Annadurai Duraisamy. "Ecological Impact of Pesticides Principally Organochlorine Insecticide Endosulfan: A Review." *Universal Journal of Environmental Research and Technology* 2, no. 5 (2012): 369–76.

Negarestani, Reza. *Cyclonopedia: Complicity with Anonymous Materials.* Re.Press, 2018.

North, Morning. "Billions of pollinating insects killed on North American roads each summer: study." *CBC News,* October 20,

2015. https://www.cbc.ca/news/canada/sudbury/insect-roadkill
-study-laurentian-1.3278025.

Oddo, Egle, and Basak Senova. "Extremophiles: The Act of
Performative Habitat." *Antennae: The Journal of Nature in
Visual Culture* 53 (2021): 131–49.

Oddo, Egle. "Evolutionary Gardens and Performative Habitats."
RUUKKU–Studies in Artistic Research 16 (2021). https://doi.org
/10.22501/ruu.792130.

Olin, Laurie. "The Campus: An American Landscape." *SiteLINES:
A Journal of Place* 8, no. 2 (2013): 3–10.

Olla, Akin. "This heatwave is a reminder that grass lawns are
terrible for the environment." *The Guardian*, July 31, 2022.
https://www.theguardian.com/commentisfree/2022/jul/31/this
-heatwave-is-a-reminder-that-grass-lawns-are-terrible-for-the
-environment.

Open Green Map. "Liz Christy Urban Community Garden, NYC."
Accessed October 10, 2024. https://www.opengreenmap.org/
greenmap/northeast-permaculture-and-ecourban-agriculture
-map/liz-christy-urban-community-garden-nyc-5905.

Overall, Mario, and Dan Hagedorn. *The 100 Hour War: The
Conflict between Honduras and El Salvador in July 1969*. Helion
& Company Limited, 2017.

Pe'er, G., et al. "EU agricultural reform fails on biodiversity."
Science 344 (2014): 1090–1092.

Pollan, Michael. *The Botany of Desire: A Plant's Eye View of the
World*. Random House Trade Paperbacks, 2002.

Ransomes, Sims & Jefferies Limited. "A great Ransomes tradition :
150 years of grasscutting technology 1832-1982." Ransomes, Sims
& Jefferies, 1982.

Reed, Eric, and Kirk O'Neil. "Super Bowl Revenue: How Much
Does the Big Game Generate?" *The Street*, February 11, 2022.

"Rethinking Highway Medians." *The Columbus Dispatch*,
September 6, 2006. https://www.dispatch.com/story/news

/technology/2007/09/04/rethinking-highway-medians
/23750620007/.

Rice, Stanly. *Green Planet: How Plants Keep the Earth Alive.*
Rutgers University Press, 2009.

Richards, Anthony. *The True Story of the Christmas Truce.*
Greenhill Books, 2021.

Robbins, Paul. *Lawn People: How Grasses, Weeds, and Chemicals
Make Us Who We Are.* Temple University Press, 2007.

Robinson, William. *The Wild Garden.* Workman Publishing, 1870.

Rogers, Elizabeth Barlow. *Saving Central Park.* Knopf, 2018.

Román-Palacios, Cristian, and John J. Wiens. "Recent responses
to climate change reveal the drivers of species extinction and
survival," *PNAS* 117, no. 8 (2020): 4211–17.

Rosen, Julia. "GMO Grass Is Creeping across Oregon." *High
Country News*, June 25, 2018.

Ruff, Allan R. *Arcadian Visions: Pastoral Influences on Poetry,
Painting and the Design of Landscape.* Windgather Press, 2015.

Saberi, Roxana. "England has its driest July in almost 90 years
as Europe swelters through historic heat waves," *CBS News*,
August 11, 2022. https://www.cbsnews.com/news/uk-driest
-july-since-1935-europe-historic-heat-waves-drought
-wildfires/.

Sandilands, Catriona, and Bruce Erickson. *Queer Ecologies: Sex,
Nature, Politics, Desire* Indiana University Press, 2010.

Sawyers, Harry, and Gregory Han. "Why We Don't Recommend
Artificial Grass for Most People." *New York Times,* July 9, 2021.
https://www.nytimes.com/wirecutter/reviews/best-artificial
-grass/.

Schiller, Friedrich. "On Naïve and Sentimental Poetry." *The Schiller
Institute*, 1795. Translated by William F. Wertz, Jr. https://
archive.schillerinstitute.com/transl/Schiller_essays/naive
_sentimental-1.html.

Scott, Frank Jesup. *The Art of Beautifying Suburban Home Grounds of Small Extent*. Appleton, 1870.

Seraphin, Bruno. "Rewilding, 'the Hoop,' and Settler Apocalypse." *The Trumpeter: Journal of Ecosophy*, 32.2 (2016): 126–46.

Seymour, Nicole. *Strange Natures*. University of Illinois Press, 2013.

Sheldrake, Merlin. *Entangled Life: How Fungi Make Our Worlds, Change Our Minds & Shape Our Futures*. Random House, 2021.

Shiva, Vandana. *Biopiracy: the plunder of nature and knowledge*. North Atlantic Books, 2016.

Simard, Suzanne. *Finding The Mother Tree*. (New York: Alfred A. Knopf, 2021)

Speller, Marcus, Luke Moore, Pete Donaldson, and Jim Campbell. *The Football Ramble*. Random House, 2016.

Steinberg, Ted. *American Green: The Obsessive Quest for the Perfect Lawn*. W.W. Norton & Company, 2006.

Stoppard, Tom. *Arcadia: A Play in Two Acts*. Samuel French, 1993.

Stromberg, Mark, Jeffrey Corbin, and Carla D'antonio. *California Grasslands: Ecology and Management*. University of California Press, 2007.

Surendran Kandasamy. "Unlocking the Potential: Transforming Highway Median Spaces into Productive Farming Land." LinkedIn, July 17, 2023. https://www.linkedin.com/pulse/unlocking-potential-transforming-highway-median-spaces-kandasamy/.

Suttie, J.M., S.G. Reynolds, and C. Batello. *Grasslands of the World*. FAO, 2005.

Teyssot, G. *The American Lawn*. Princeton Architectural Press with Canadian Centre For Architecture, 2001.

Toland, Alexandra, Jay S. Noller, and Gerd Wessolek. *Field to Palette: Dialogues on Soil and Art in the Anthropocene*. CRC Press, 2018.

Tsing, Anna. "Contaminated Diversity in 'Slow Disturbance':
Potential Collaborators for a Liveable Earth." In *Why Do We
Value Diversity? Biocultural Diversity in a Global Context*,
edited by Gary Martin, Diana Mincyte, and Ursula Münster.
RCC Perspectives, 2012.

Turner, Victor, Roger D Abrahams, and Alfred Harris. *The Ritual
Process*. Routledge, 2017.

Vadukul, Alex. "When an Ax-Wielding Mob Leveled a Gay
Cruising Spot as the Police Watched." *New York Times,* July 2,
2019. https://www.nytimes.com/2019/07/02/automobiles/when
-an-ax-wielding-mob-leveled-a-gay-cruising-spot-as-police
-watched.html.

Vanek, Corina. "Super Bowl 57: $1.3B, 103K out-of-state visitors
to Arizona. And more." *The Arizona Republic*, October 4, 2023.
https://www.azcentral.com/story/money/business/economy
/2023/10/04/super-bowl-57-brought-1-3-billion-in-economic
-impact-to-arizona/71063201007/.

Van Gennep, Arnold. *The Rites of Passage*. Routledge, 1960.

Viveiros de Castro, Eduardo. "Zeno and the Art of Anthropology:
Of Lies, Beliefs, Paradoxes, and Other Truths." *Common
Knowledge* 17, no. 1 (2011): 128–45.

Wall Kimmerer, Robin. "Building Good Soil." In *What Kind of
Ancestor Do You Want to Be?*, edited by John Hausdoerffer,
Brooke Parry Hecht, Melissa K. Nelson, and Katherine Kassouf
Cummings. University of Chicago Press, 2021.

Wandersee, James H., and Elizabeth E. Schussler. "Preventing
Plant Blindness." In *The American Biology Teacher* 61, no. 2
(1999): 82–86.

Weinberger, Lois, and Komm.Rat.-Dr.-Hans-Klocker-Und-Dr.-
Wolfgang Klocker-Stiftung, *Lois Weinberger* (Ostfildern: Hatje
Cantz, 2014).

Weinfuss, Josh. "Behold the Super Bowl's experimental golf grass years in the making." ESPN.com. February 13, 2023. https://www.espn.com/nfl/story/_/id/35625308/super-bowl-2023-experimental-grass.

White, Lynn. "The Historical Roots of Our Ecologic Crisis." *Science* 155, no. 3767 (1967): 1203–07.

Williams, Jack. *Cricket and England*. Routledge, 2012.

Wohlleben, Peter and Billinghurst, Jane. *The Hidden Life of Trees: What They Feel, How They Communicate: Discoveries from a Secret World*. Vancouver: Greystone Books, 2016.

Zajko, Vanda and Hoyle, Hoyle. *A Handbook to the Reception of Classical Mythology*. New York: John Wiley & Sons, Inc. 2017. https://www.wiley.com/en-us/A+Handbook+to+the+Reception+of+Classical+Mythology-p-9781444339604/

Zee, Renate Helena Hoyle. "'We Hate the Headscarf': Can Women Find Freedom in Tehran's Female-Only Parks?" *The Guardian*, August 9, 2017. https://www.theguardian.com/cities/2017/aug/09/women-only-parks-tehran-iran-segregated-outside-spaces .

Žižek, Slavoj. *Violence: Six Sideways Reflections*. Picador, 2008.

Žižek, Slavoj. *Event: A Philosophical Journey through a Concept*. Melville House, 2014.

Žižek, Slavoj. *The Sublime Object of Ideology*. Verso, 2019.

NOTES

Preface

1 Jonathan Fraser Light, *The Cultural Encyclopedia of Baseball* (London: McFarland and Company, 2005).

2 Tyne Morgan, "Artificial Turf Made With Soybeans is Growing in Popularity, Now on Display at the San Diego Zoo," *The Scoop*, March 14, 2023, accessed January 20, 2024, https://www.thedailyscoop.com/news/retail-industry /artificial-turf-made-soybeans-growing-popularity-now -display-san-diego-zoo#:~:text=The%20artificial%20turf%20 market%20is.

3 Jared Firestone, "Artificial Turf Cancer Lawsuit: Toxic Compounds May be Linked to Cancer," Expert Institute, accessed September 8, 2022, https://www.expertinstitute .com/resources/insights/artificial-turf-cancer-lawsuit-toxic -compounds-may-be-linked-to-cancer/.

4 Harry Sawyers and Gregory Han, "Why We Don't Recommend Artificial Grass for Most People," *New York Times*, July 9, 2021, accessed September 24, 2022, https://www .nytimes.com/wirecutter/reviews/best-artificial-grass/.

5 Francie Diep, "Lawns vs. crops in the continental U.S.," *Scienceline*, September 19, 2011, accessed March 5, 2023, https://scienceline.org/2011/07/lawns-vs-crops-in-the-continental-u-s/.

6 Tom Fort, *The Grass is Greener* (New York: Harper Collins, 2008); Georges Teyssot, ed., *The American Lawn* (New Jersey: Princeton Architectural Press with Canadian Centre For Architecture, Montréal, 2001).

7 Roxana Saberi, "England has its Driest July in Almost 90 years as Europe Swelters through Historic Heat Waves," *CBS News*, August 11, 2022, accessed July 31, 2023, https://www.cbsnews.com/news/uk-driest-july-since-1935-europe-historic-heat-waves-drought-wildfires/; Akin Olla, "This heatwave is a reminder that grass lawns are terrible for the environment," *The Guardian*, July 31, 2022, accessed March 5, 2023, https://www.theguardian.com/commentisfree/2022/jul/31/this-heatwave-is-a-reminder-that-grass-lawns-are-terrible-for-the-environment.

8 Kiana Dickson, "What is the Anti-lawn Movement? Expert Opinions and How You Can Join in with the Change," *Better Homes and Gardens*, September 7, 2022, accessed March 5, 2023, https://www.homesandgardens.com/gardens/what-is-the-anti-lawn-movement.

9 Virginia Scott Jenkins, *The Lawn: A History of an American Obsession* (Washington DC: Smithsonian Institution Press, 1999).

10 Fort, *The Grass is Greener*; Teyssot, *The American lawn*.

11 Paul Robbins, *Lawn People: How Grasses, Weeds, and Chemicals Make Us Who We Are* (Philadelphia: Temple University Press, 2007). Steinberg, Ted. *American Green: The*

Obsessive Quest for the Perfect Lawn (New York: W.W. Norton, 2006).

12 Jonathan Cane, *Civilising Grass* (Johannesburg: Wits University Press, 2019).

13 F Herbert Bormann, Balmori, D., and Geballe, G. T., *Redesigning the American Lawn: A Search for Environmental Harmony* (New Haven: Yale University Press, 2001).

14 For an overview of the ontological turn in the humanities, see Paolo Heywood, "Anthropology and What There Is: Reflections on 'Ontology,'" *Cambridge Journal of Anthropology* 30, no. 1 (2012): 143–51; Paolo Heywood, "The Ontological Turn," in *Cambridge Encyclopedia of Anthropology* (Cambridge: Cambridge University Press, 2017); Eduardo Kohn, "Anthropology of Ontologies," *Annual Review of Anthropology* 44, no. 1 (2015): 311–27, https://doi .org/10.11 46/annurev-anthro-102214-014127; and Eduardo Viveiros de Castro, "Zeno and the Art of Anthropology: Of Lies, Beliefs, Paradoxes, and Other Truths," *Common Knowledge* 17, no. 1 (2011): 128–45.

15 Giovanni Aloi, ed., "Why Look at Plants?," special issue, *Antennae: The Journal of Nature in Visual Culture,* no. 17 (2011); Giovanni Aloi, ed., "Beyond Morphology," special issue, *Antennae: The Journal of Nature in Visual Culture,* no. 18 (2011); Michael Marder, *Plant-Thinking: A Philosophy of Vegetal Life* (New York: Columbia University Press, 2013); Matthew Hall, *Plants as Persons: A Philosophical Botany* (Albany: SUNY Press, 2011); Monica Gagliano, *Thus Spoke the Plant: A Remarkable Journey of Groundbreaking Scientific Discoveries and Personal Encounters with Plants* (London: Penguin, 2018); Stefano Mancuso and Alessandra Viola, *Brilliant Green: The Surprising History and Science of Plant*

Intelligence (Washington, D.C.: Island Press, 2015); Giovanni Aloi, ed. and principal au., *Why Look at Plants? The Botanical Emergence in Contemporary Art* (Leiden, Netherlands: Brill, 2018).

16 Giovanni Aloi and Susan McHugh, *Posthumanism in Art and Science* (New York: Columbia University Press, 2021); Braidotti, Rosi, *The Posthuman* (Cambridge: *Polity,* 2013); Karen Barad, *Meeting the Universe Halfway: Physics and the Entanglement of Matter and Meaning* (Durham, N.C.: Duke University Press, 2007); *Donn Haraway, Simians, Cyborgs, and Women: The Reinvention of Nature* (New York: Routledge, 1991); N. Katherine Hayles, "Unfinished Work: From Cyborg to Cognisphere," *Theory, Culture and Society* 23, no. 7–8 (2006), 159–66.

17 Robin MacKay, Luke Pendrell, and James Trafford, *Speculative Aesthetics* (Falmouth: Urbanomic, 2014); Levi Bryant, Nick Srnicek, and Graham Harman, *The Speculative Turn* (Melbourne: RePress, 2011).

18 Graham Harman, *Third Table,* part of the series *100 Notes, 100 Thoughts: dOCUMENTA* (Berlin: Hatje Cantz, 2012); Graham Harman, "Art and OOObjecthood," in C. Cox, J. Jaskey, and S. Malik, eds., *Realism Materialism Art* (Berlin: Sternberg, 2015), 97–122; Graham Harman, *The Quadruple Object* (Ropley: Zero Press, 2011); Timothy Morton, *Ecology without Nature: Rethinking Environmental Aesthetics* (Cambridge: Harvard University Press, 2007); Timothy Morton, *Hyperobjects: Philosophy and Ecology after the End of the World* (Minneapolis: University of Minnesota Press, 2017).

19 Robin Wall Kimmerer, *Braiding Sweetgrass: Indigenous Wisdom, Scientific Knowledge, and the Teachings of Plants*

(Minneapolis: Milkweed Press, 2014); Eduardo Kohn, *How Forests Think: Toward an Anthropology Beyond the Human* (Berkeley: University of California Press, 2015).

20 Catriona Sandilands and Bruce Erickson, *Queer Ecologies: Sex, Nature, Politics, Desire* (Bloomington: Indiana University Press, 2010); Nicole Seymour, *Strange Natures* (Champaign: University of Illinois Press, 2013).

21 Ted Steinberg, *American Green: The Obsessive Quest for the Perfect Lawn* (Washington: W. W. Norton & Company, 2007).

22 Michel Foucault, *Discipline and Punish: The Birth of the Prison*, trans. Alan Sheridan (1975; repr., New York: Vintage Books, 1995). https://www.google.com/books/edition/Discipline_and_Punish/6rfP0H5TSmYC?hl=en&gbpv=1.

23 Reza Negarestani, *Cyclonopedia: Complicity with Anonymous Materials* (Melbourne: Re.Press, 2020), 85.

Chapter 1

1 P.A. Clarke, "The Aboriginal Australian cosmic landscape, Part 1: The Ethnobotany of the Skyworld," *Journal of Astronomical History and Heritage*, 2014, 17(3), 307–335.

2 Duane Hamacher and Kirsten Banks, "The Planets in Indigenous Australian Traditions," in *Oxford Research Encyclopedia of Planetary Science*, ed. Peter Read (Oxford: Oxford University Press, 2018).

3 Rosalyn LaPier, "Indigenous Knowledge, Grasslands and Bison," The American Buffalo: A Film by Ken Burns, PBS, accessed July 1, 2024, https://www.pbs.org/kenburns/the

-american-buffalo/indigenous-knowledge-grasslands-and
-bison.

4 Kat Anderson, *Tending the Wild: Native American Knowledge
and the Management of California's Natural Resources*
(Berkeley: University Of California Press, 2013).

5 Patricia Ann Lynch and Jeremy Roberts, *African Mythology A
to Z* (New York: Chelsea House, 2010), 36.

6 Robin Wall Kimmerer, *Braiding Sweetgrass: Indigenous
Wisdom, Scientific Knowledge, and the Teachings of Plants*
(Minneapolis: Milkweed Editions, 2013).

7 Kimmerer, *Braiding Sweetgrass*, 383.

8 Illinois Prairies, accessed January 10, 2024, https://dnr.illinois
.gov/education/atoz/ilprairies.html.

9 J. M. Suttie, S. G. Reynolds, and C. Batello, *Grasslands of the
World.* (Rome: FAO, 2005).

10 Pe'er, G., et al. 2014. "EU agricultural reform fails on
biodiversity," *Science* 344: 1090–1092.

11 J. Bengtsson, et al. "Grasslands—more important for
ecosystem services than you might think," *Ecosphere* 10, no. 2
(February 2019): e02582. https://doi.org/10.1002/ecs2.2582.

12 Ole Jørgen Benedictow, *The Complete History of the Black
Death* (Martlesham: Boydell & Brewer, 2021).

13 Lynn White, "The Historical Roots of Our Ecologic Crisis,"
Science, 1967, *155*(3767), 1203–1207.

14 Amitav Ghosh, *The Great Derangement: Climate Change and
the Unthinkable* (Chicago: The University of Chicago Press,
2016).

15 Vandana Shiva, *Biopiracy: The Plunder of Nature and
Knowledge* (Berkeley: North Atlantic Books, 2016).

16 James H. Wandersee and Elisabeth E. Schussler, "Preventing Plant Blindness," *The American Biology Teacher* 61, no. 2 (February 1999): 84.

17 Peter Wohlleben and Jane Billinghurst, *The Hidden Life of Trees: What They Feel, How They Communicate: Discoveries from a Secret World* (Vancouver: Greystone Books, 2016); Suzanne Simard, *Finding The Mother Tree* (New York: Alfred A. Knopf, 2021).

18 Suttie Reynolds, and Batello, *Grasslands of the World*, Rome: Food and Agriculture Organization of the United Nations, 2005)

19 Stanly Rice, *Green Planet: How Plants Keep the Earth Alive* (New Brunswick: Rutgers University Press, 2009).

20 Robert Kourik and R. Creasy, *Roots Demystified: Designing and Maintaining Your Edible Landscape Naturally* (Emmaus: Metamorphic Press, 1986).

21 Stephen Harris, *Grasses* (London: Reaktion Books, 2014), 155.

22 Harris, *Grasses*, 129.

23 Jenny Kendler, "An Open Book of Grass" in *Why Look at Plants? The Botanical Emergence in Contemporary Art*, ed. Giovanni Aloi (Leiden: Brill), 61–66.

24 Gilles Deleuze and Félix Guattari, *A Thousand Plateaus: Capitalism and Schizophrenia* (B. Massumi, Trans.). (London: Bloomsbury, 2017).

Chapter 2

1 Dante Alighieri and M. Musa, *Dante's Inferno: The Indiana Critical Edition* (Bloomington: Indiana University Press, 1995), 3.

2 Alighieri and Musa, 4.

3 Alighieri and Musa.

4 Vanda Zajko and Helena Hoyle, *A Handbook to the Reception of Classical Mythology* (New York: John Wiley & Sons, Inc. 2017).

5 Zajko and Hoyle, 305.

6 A.T. Murray, *Homer, The Odyssey with an English Translation* (Cambridge: Harvard University Press, 1991), 560–565.

7 Anthony Huxley, *An Illustrated History of Gardening* (New York and London: Paddington Press LTD, 1978).

8 Arie Graafland, "The Dance of Versailles: Nature, Circe, and the Garden," in *Earth Perfect: Nature, Utopia, and the Garden* (London: Black Dog Publishing, 2012), 84–103

9 Edmund Burke, *A Philosophical Inquiry into the Origin of Our Ideas* (New York: Simon and Schuster, 2012).

10 Ransomes, Sims & Jefferies, *150 Years of Grasscutting Technology, 1832–1982* (Ipswich, England: 1982), 1. 132.

11 Mary Herring, "From Sheep to Robots: The History of Lawn Mowers," Iron Solutions, 2020, accessed January 8, 2024, https://ironsolutions.com/the-history-of-lawn-mowers/#:~:text=Steam%2Dpowered%20lawn%20mowers%20 appeared.

12 Fort, 10–17.

13 Steinberg, 10–13.

14 Olivia Milman, "Tree-mendous news: noisy gas-powered leaf blowers banned in Washington DC," *The Guardian*, January 5, 2022, accessed January 5, 2024, https://www.theguardian.com/us-news/2022/jan/05/gas-leaf-blowers-banned-washington-dc.

15 James Fellows, "Leaf blower Legislation: The *WaPo* and the *Current* Weigh In," *The Atlantic,* January 2016, accessed February 3, 2023, https://www.theatlantic.com/national/archive/2016/01/leafblower-legislation-local-press/625260/.

16 Andy Gregory, "Leaf blowers contributing to 'insect armageddon' and should be avoided, German government warns," *The Independent*, November 19, 2019, accessed January 8, 2024, https://www.independent.co.uk/climate-change/news/leaf-blowers-insect-noise-pollution-germany-gardening-a9208366.html.

17 *Blue Velvet*, directed by David Lynch (Warner Brothers, 1986).

Chapter 3

1 Council on Foreign Relations, "Global Conflict Tracker l Council on Foreign Relations," Global Conflict Tracker, 2023, accessed January 20, 2024, https://www.cfr.org/global-conflict-tracker.

2 Anna Duggal and Ali Marium, "Why do more than 800 million people live in hunger?" Al Jazeera, May 23, 2023, accessed January 17, 2024, https://www.aljazeera.com/news/2023/5/28/why-is-global-hunger-on-the-rise-2#:~:text=Hunger%20levels%20are%20rising%20around.

3 Cristian Román-Palacios and John J. Wiens, "Recent responses to climate change reveal the drivers of species extinction and survival" *PNAS* 117, no. 8, 2020. https://www.pnas.org/doi/10.1073/pnas.1913007117.

4 IPCC. *Summary for Policymakers of IPCC Special Report on Global Warming of 1.5°C approved by governments — IPCC.*

Ipcc.ch; IPCC. October 8, 2018, accessed January 22, 2024, https://www.ipcc.ch/2018/10/08/summary-for-policymakers -of-ipcc-special-report-on-global-warming-of-1-5c-approved -by-governments/.

5 Racheal Fest. "Culture and Neoliberalism: Raymond Williams, Friedrich Hayek, and the New Legacy of the Cultural Turn." Mediations 34.2 (Spring 2021) 9-32. www.mediationsjournal .org/articles/culture-and-neoliberalism, accessed January 4, 2024,

6 Tyler Lauletta, "Eagles and Chiefs players both agreed that the Super Bowl's $800,000 turf was terrible: "It was like playing on a water park," *Business Insider*, February 13, 2023, accessed January 17, 2024, https://www .businessinsider.com/eagles-chiefs-super-bowl-field-turf -bad-slip-water-park-2023-2#:~:text=Eagles%20and%20 Chiefs%20players%20both.

7 Jeffrey Cluff, "Arizona Cardinals' field slammed again as 'atrocious,' 'deplorable,' 'embarrassing,'" *The Arizona Republic*, August 14, 2023, https://www.azcentral.com/story/sports/nfl/ cardinals/2023/08/14/arizona-cardinals-field-turf-nfl-playing -surface-slammed-again/70589468007/.

8 Chris Davies, "The High-Tech 2-Year Story Behind Super Bowl 2023's Disastrous Turf," *SlashGear*, February 13, 2023, accessed January 10, 2024, https://www.slashgear.com /1197410/the-high-tech-2-year-story-behind-super-bowl -2023s-disastrous-turf/.

9 Josh Weinfuss, "Behold the Super Bowl's experimental golf grass years in the making," ESPN.com, February 13, 2023, accessed January 15, 2023, https://www.espn.com/ nfl/story/_/id/35625308/super-bowl-2023-experimental- grass.

10 Jeffrey Cluff, 2023; Lorenzo Reyes, "'The Sodfather' George Toma explains why Super Bowl 57 field was so slippery," *USA Today*, February 28, 2023, accessed January 8, 2024, https://www.usatoday.com/story/sports/nfl/super-bowl/2023/02/28/super-bowl-57-slippery-field-conditions-explained/11368392002/.

11 Veronika Bondarenko, Eric Reed, and Kirk O'Neil, "Super Bowl Revenue: How Much Does the Big Game Generate?" *The Street*, February 11, 2022, accessed January 12, 2024, https://www.thestreet.com/lifestyle/sports/super-bowl-revenue.

12 Corina Vanek, "Super Bowl 57: $1.3B, 103K out-of-state visitors to Arizona. And more.," *The Arizona Republic*, October 4, 2023, accessed January 8, 2024, https://www.azcentral.com/story/money/business/economy/2023/10/04/super-bowl-57-brought-1-3-billion-in-economic-impact-to-arizona/71063201007/.

13 Veronika Bondarenko, Eric Reed, and Kirk O'Neil. (2022, February 11).

14 Ibid.

15 Carla Johnson, "What does the science say about the grass vs. turf debate in sports?" *AP News*, October 9, 2023, accessed January 15, 2024, https://apnews.com/article/nfl-aaron-rodgers-achilles-grass-artificial-turf-79212f5443cd2a0d30fe8c9d981b13c0.

16 Dave Anderson, "It's Time to Clean Up the War Vocabulary in Sports," *New York Times,* March 23, 2003, accessed January 20, 2024, https://www.proquest.com/newspapers/time-clean-up-war-vocabulary-sports/docview/432335574/se-2.

17 Martin Meredith, *Mandela* (New York: Simon and Schuster, 2010).

18 Mario Overall and Dan Hagedorn, *The 100 Hour War: The Conflict between Honduras and El Salvador in July 1969* (Warwick: Helion & Company Limited, 2017).

19 Anthony Richards, *The True Story of the Christmas Truce* (Barnsley: Greenhill Books, 2021).

20 Emanuel Leite Junior and C. Rodrigues, *China, Football, and Development* (New York: Taylor & Francis, 2023).

21 Marcus Speller, Luke Moore, Pete Donaldson, Jim Campbell, *The Football Ramble* (London: Random House, 2016).

22 Gregory M. Reichberg, *Thomas Aquinas on War and Peace* (Cambridge: Cambridge University Press, 2018).

23 Jay Coakley, *Handbook of Sports Studies* (Newcastle upon Tyne: Sage, 2010).

24 Graham Curry and Eric Dunning, *Association Football* (London: Routledge, 2015).

25 Mark. F. Bernstein, *Football: The Ivy League Origins of an American Obsession* (Philadelphia: University Of Pennsylvania Press, 2001).

26 Jack Williams, *Cricket and England* (London: Routledge, 2012).

27 Haslam, S Alexander. Fransen, Katrien. and Boen, Filip. *The New Psychology of Sport and Exercise* (London: SAGE, 2020).

28 Slavoj Žižek, *Violence: Six Sideways Reflections* (London: Picado, 2008); Slavoj Žižek, *Event: A Philosophical Journey through a Concept* (Melville House, 2014); Slavoj Žižek, *The Sublime Object of Ideology* (New York: Verso, 2019).

29 Marcel Merleau-Ponty, *The Structure of Behavior*, trans. Fischer (London: Methuen, 1965), 168.

30 Chantelle Billson, "Study reveals sports most notorious for homophobic and transphobic abuse," *PinkNews*, April 22, 2023, accessed January 10, 2023, https://www.thepinknews .com/2023/04/22/sports-football-cricket-homophobic -transphobic-abuse/.

Chapter 4

1 Renate van der Zee, "'We Hate the Headscarf': Can Women Find Freedom in Tehran's Female-Only Parks?" *The Guardian*, August 9, 2017, accessed February 8, 2024, https://www .theguardian.com/cities/2017/aug/09/women-only-parks -tehran-iran-segregated-outside-spaces.

2 Zee, "We Hate the Headscarf."

3 Reza Arjmand, *Public Urban Space, Gender and Segregation* (London: Routledge, 2016).

4 Judith Butler, *Gender Trouble: Feminism and the Subversion of Identity* (New York: Routledge, 1999).

5 Michel Foucault, *Discipline and Punish: The Birth of the Prison* (New York: Vintage Books, 1975).

6 Alex Vadukul, "When an Ax-Wielding Mob Leveled a Gay Cruising Spot as the Police Watched," *New York Times,* July 2, 2019, accessed February 10, 2024, https://www.nytimes.com /2019/07/02/automobiles/when-an-ax-wielding-mob-leveled -a-gay-cruising-spot-as-police-watched.html.

7 Catriona Mortimer-Sandilands and Bruce Erickson, eds., *Queer Ecologies: Sex, Nature, Politics, Desire* (Bloomington, Ind.: Indiana University Press, 2010).

8 Mortimer-Sandilands and Erickson, *Queer Ecologies*, 4.

9 Frank Jesup Scott, *The Art of Beautifying Suburban Home Grounds of Small Extent* (New York: Appleton, 1870), 61.

10 Hope Lourie Killcoyne and Mary Lee Majno, *The Lost Village of Central Park* (New York: Silver Moon Press, 1999).

11 Morrison H. Heckscher, *Creating Central Park* (New York: Metropolitan Museum of Art, 2008).

12 Allan R. Ruff, *Arcadian Visions: Pastoral Influences on Poetry, Painting and the Design of Landscape* (Oxford: Windgather Press, An Imprint Of Oxbow Books; Havertown, Pa, 2015).

13 Elizabeth Barlow Rogers, *Saving Central Park* (New York: Knopf, 2018).

14 Wilfred Chan, "Central Park Warned a Festival Would Ruin Its Lawn. New York Went Ahead with It," *The Guardian*, October 6, 2023, accessed February 19, 2024, https://www .theguardian.com/us-news/2023/oct/06/central-park-great -lawn-global-citizen-festival.

15 Theodore Steinberg, *American Green: The Obsessive Quest for the Perfect Lawn* (New York: W.W. Norton, 2006).

16 Alan Feuer, "Settlement on Use of Central Park's Great Lawn," *New York Times*, January 8, 2008, accessed February 16, 2024, https://www.nytimes.com/2008/01/08/nyregion/09cnd-lawn .html.

Chapter 5

1 The Record-Currier Editors, "Anarchist Golfing Association Trashes Grass-Seed Research Facility," *The Record-Currier,*

December 20, 2001, https://www.recordcourier.com/news
/2001/dec/20/anarchist-golfing-association-trashes-grass-seed
-r/.

2   The Record-Currier Editors, "Anarchist Golfing."

3   Mike Kenna,"A Turfgrass Timeline: The History of Creeping
Bentgrass Breeding," USGA, 2024, accessed March 10, 2024,
https://www.usga.org/content/usga/home-page/course-care
/green-section-record/62/issue-01/a-turfgrass-timeline--the
-history-of-creeping-bentgrass-breeding.html.

4   Julia Rosen, "GMO Grass Is Creeping across Oregon," *High
Country News*, June 25, 2018, accessed March 10, 2024,
https://www.hcn.org/issues/50-11/plants-genetically-modified
-grass-creeps-across-eastern-oregon/.

5   Friedrich Schiller, "On Naïve and Sentimental Poetry," *The
Schiller Institute*, 1795, trans. William F. Wertz, Jr.

6   Louise A. Mozingo, *Pastoral Capitalism* (Cambridge: MIT
Press, 2016), 222.

7   C. A. Kearns and L. Prior, "Toxic greens: a preliminary
study on pesticide usage on golf courses in Northern Ireland
and potential risks to golfers and the environment," *WIT
Transactions on The Built Environment* 134, 2013, 173–182.

8   Kristi Dosh, "Golfers Make Better Business Executives,"
Forbes, May 16, 2016, accessed February 28, 2024, https://
www.forbes.com/sites/kristidosh/2016/05/16/golfers-make
-better-business-executives/?sh=28436750b4a5.

9   National Golf Foundation, "Golf Research and Industry Data,"
ngf.org, 2023, accessed February 28, 2024, https://www.ngf
.org/golf-industry-research/.

10   Thomas Kingsley, "Climate Activists Fill Golf Course Holes
with Cement," *The Independent,* August 14, 2022, accessed

February 29, 2024, https://www.independent.co.uk/climate
-change/news/climate-activist-france-water-ban-b2144729
.html.

11 Sam Jones, "Extinction Rebellion Plugs Holes on 10 Spanish
Golf Courses in Water Protest," *The Guardian*, July 3, 2023,
accessed February 29, 2024, https://www.theguardian.com
/world/2023/jul/03/extinction-rebellion-plugs-holes-on-10
-spanish-golf-courses-in-water-protest.

12 Ben Adler, "The Case against Golf," *The Guardian,* June 14,
2007, accessed February 29, 2024, https://www.theguardian
.com/commentisfree/2007/jun/14/thecaseagainstgolf.

13 Neelambari Vivekanandhan and Annadurai Duraisamy,
"Ecological Impact of Pesticides Principally Organochlorine
Insecticide Endosulfan: A Review," *Universal Journal of
Environmental Research and Technology* 2, no. 5: 369–376.

14 Laurie Olin, "The Campus: An American Landscape,"
SiteLINES: A Journal of Place 8, no. 2 (Spring 2013), 3–10.

Chapter 6

1 "Rethinking Highway Medians," *The Columbus Dispatch*,
September 6, 2006, accessed March 9, 2023, https://www
.dispatch.com/story/news/technology/2007/09/04/rethinking
-highway-medians/23750620007/.

2 Reza Negarestani, *Cyclonopedia: Complicity with Anonymous
Materials* (Melbourne: Re.Press, 2018), 85.

3 Morton, Timothy. Hyperobjects: Philosophy and Ecology
after the End of the World. (Minneapolis: University Of
Minnesota Press, 2013)

4   Michael Pollan, *The Botany of Desire: A Plant's Eye View of the World* (New York: Random House Trade Paperbacks, 2002).

5   Merlin Sheldrake, *Entangled Life: How Fungi Make Our Worlds, Change Our Minds & Shape Our Futures* (S.L.: Random House, 2021).

6   Gilles Deleuze and Félix Guattari, *A Thousand Plateaus: Capitalism and Schizophrenia* (B. Massumi, Trans.), (London: Bloomsbury: 1980, 2017).

7   David Gardner and Thomas deHaas, "No Mow May in Full Swing.........So What's Wrong with 'No Mow May'?" *Buckeye Yard and Garden*, May 12, 2023, accessed February 17, 2024, https://bygl.osu.edu/node/2137.

8   Arnold van Gennep, *The Rites of Passage* (London: Routledge, 1960).

9   Victor Turner, Roger D. Abrahams, and Alfred Harris, *The Ritual Process* (London: Routledge, 2017).

10 Terrill L. Gibson, *The Liminal and the Luminescent* (Wipf and Stock Publishers, 2021).

11 Homi K. Bhabha, *The Location of Culture* (London: Routledge, 2012).

12 Folukei Fejola, "Why I Say 'Decolonisation Is Impossible,'" *Foluke's African Skies,* December 17, 2019, accessed January 3, 2024, https://folukeafrica.com/why-i-say-decolonisation-is-impossible/#:~:text=Dispossession%20always%20serves%20the%20purpose.

13 J. Jennifer Foote, "Radical environmentalists are honing their militant tactics and gaining followers," *Newsweek,* 115, 1990, 24.

14 Dolly Jørgensen, "Rethinking Rewilding," *Geoforum,* 65 2015, 482–488

15 William Cronon, "The Trouble with Wilderness: Or, Getting Back to the Wrong Nature," *Environmental History,* 1.1 (1996), 7–28.

16 Bruno Seraphin, "Rewilding, 'the Hoop,' and Settler Apocalypse," *The Trumpeter: Journal of Ecosophy*, 32.2 (2016), 126–46; 'Paiutes and Shoshone', 447–78.

17 Surendran Kandasamy, "Unlocking the Potential: Transforming Highway Median Spaces into Productive Farming Land," LinkedIn, July 17, 2023, accessed March 29, 2024, https://www.linkedin.com/pulse/unlocking-potential -transforming-highway-median-spaces-kandasamy/ ?trackingId=NiXkapV8Rk%2BGOiq99XjFvg%3D%3D.

18 Troy Bishop, "Median Grazing: Crazy Enough to Work?" *Country Folks*, December 20, 2013, accessed March 29, 2024, https://countryfolks.com/median-grazing-crazy-enough-to -work/.

19 Damian Carrington, "Car 'Splatometer' Tests Reveal Huge Decline in Number of Insects," *The Guardian*, February 12, 2020, accessed March 29, 2024, https://www.theguardian.com /environment/2020/feb/12/car-splatometer-tests-reveal-huge -decline-number-insects.

20 Morning North, "Billions of pollinating insects killed on North American roads each summer: study," *CBC News*, 2015, accessed March 27, 2024, https://www.cbc.ca/news/canada/ sudbury/insect-roadkill-study-laurentian-1.3278025.

21 Alistair Bonnett, *Off the Map* (London: Arum Press Ltd, 2014).

22 Benjamin Heim Shepard, *Sustainable Urbanism and Direct Action* (Rowman & Littlefield, 2014).

23 Mark Levitan and Susan Wieler, "Poverty in New York City, 1969–99: The Influence of Demographic Change, Income Growth, and Income Inequality," *SSRN Electronic Journal*, July 2008, https://doi.org/10.2139/ssrn.1185602.

24 "Liz Christy Urban Community Garden," Open Green Map, accessed March 29, 2024, https://www.opengreenmap.org/greenmap/northeast-permaculture-and-ecourban-agriculture-map/liz-christy-urban-community-garden-nyc-5905.

25 Cecilia Gowdy-Wygant, *Cultivating Victory: The Women's Land Army and the Victory Garden Movement* (Pittsburgh: University of Pittsburgh Press, 2013).

26 Alexandra Toland, Jay S. Noller, and Gerd Wessolek, *Field to Palette: Dialogues on Soil and Art in the Anthropocene* (CRC Press, 2018).

27 Deleuze and Guattari, (2017).

28 "Future Farmers: About," futurefarmers.com, 2022, accessed February 12, 2024, https://www.futurefarmers.com/about.

29 Will Allen and Fritz Haeg, *Edible Estates: Attack on the Front Lawn: A Project by Fritz Haeg* (Editorial: New York, NY: Metropolis Books, 2010).

30 Valerie Padilla Carroll, *Who Gets to Go Back-To-The-Land?* (U of Nebraska Press, 2020).

31 Lana Miran, "How Artist T'uy't'tanat-Cease Wyss Uses Ethnobotany to Reconnect Unceded Land with Indigenous Culture," *Arts Help,* September 20, 2023, accessed November 4, 2023, https://www.artshelp.com/how-artist-tuyttanat-cease-wyss-uses-ethnobotany-to-reconnect-unceded-land-with-indigenous-culture/.

32 Tom Stoppard, *Arcadia: A Play in Two Acts* (New York; London: Samuel French, 1993), 25–26.

33 Lois Weinberger and Komm.Rat Dr.-Hans-Klocker und-Dr.-Wolfgang Klocker-Stiftung, *Lois Weinberger* (Ostfildern: Hatje Cantz, 2014).

34 Jamie Grierson, "Chelsea Flower Show Gardens to Rebrand Weeds as 'Hero' Plants," *The Guardian*, April 25, 2023, accessed February 23, 2024, https://www.theguardian.com /lifeandstyle/2023/apr/25/chelsea-flower-show-gardens -rebrand-weeds-hero-plants.

35 William Robinson, *The Wild Garden* (Portland: Workman Publishing, 1870), 199

36 Alys Fowler, "Ditch Your Spade, Forget Fertiliser, Listen to the Weeds: Alys Fowler's Guide to Laid-Back Gardening," *The Guardian*, May 10, 2023, accessed February 23, 2024, https:// www.theguardian.com/lifeandstyle/2023/may/10/ditch-your -spade-forget-fertiliser-listen-to-the-weeds-alys-fowlers-guide -to-laid-back-gardening.

37 Fowler, "Ditch Your Spade."

38 Egle Oddo and Basak Senova, "Extremophiles: The Act of Performative Habitat" in *Antennae* 53, 131–149.

39 Egle Oddo, "Evolutionary Gardens and Performative Habitats," *RUUKKU—Studies in Artistic Research* 16, May 21, 2021, accessed January 12, 2024, https://www .researchcatalogue.net/view/792130/792131.

40 Lurie Gardens Stewardship. "Sustainability." Lurie Garden, n.d. https://www.luriegarden.org/about/sustainability.

41 Jo ana Kubiak, "Cutting Back on the Cut-Back," *Lurie Garden,* March 15, 2019, accessed January 23, 2024, https://www .luriegarden.org/cutting-back-on-the-cut-back/.

42 Anna Tsing, "Contaminated Diversity in 'Slow Disturbance': Potential Collaborators for a Liveable Earth," *Why Do We*

Value Diversity? Biocultural Diversity in a Global Context, eds. Gary Martin, Diana Mincytė, and Ursula Münster, RCC Perspectives 2012, no. 9, 95–97.

43 Robin Wall Kimmerer, "Building Good Soil," *What Kind of Ancestor Do You Want to Be?* eds. John Hausdoerffer, Brooke Parry Hecht, Melissa K. Nelson, and Katherine Kassouf Cummings (Chicago: University of Chicago Press, 2021), 182–184.

44 Kimmerer, 183.

45 Kimmerer, 184.

INDEX